Mastering the Ohio Graduation Test in Science

MARK JARRETT

STUART ZIMMER

JAMES KILLORAN

JARRETT PUBLISHING COMPANY

EAST COAST OFFICE
Post Office Box 1460
Ronkonkoma, NY 11779
631-981-4248

WEST COAST OFFICE
10 Folin Lane
Lafayette, CA 94549
925-906-9742

1-800-859-7679 Fax: 631-588-4722
www.jarrettpub.com

Cover photo: Corbis, Inc.

Jarrett Publishing Company
Post Office Box 1460
Ronkonkoma, New York 11779

ISBN 1-882422-84-8
Printed in the United States of America
by Malloy, Incorporated, Ann Arbor, Michigan
First Edition
10 9 8 7 6 11 10

ACKNOWLEDGMENTS

The authors would like to thank the following educators who helped review the manuscript. Their collective comments, suggestions, and recommendations have proved invaluable in preparing this book.

Lori A. Jackson
Science Teacher, Whetstone High School
Science Curriculum Writer
Columbus Public Schools
Columbus, Ohio

Tom Stork
Science Chairperson
Physics Teacher
Athens High School
Athens, Ohio

Sheri Zakarowsky, NBCT
Greater Cleveland
Educational Development Center
Cleveland State University
Teacher on Loan — Northeast Ohio Region
Cleveland, Ohio

Charles Rexer
Retired Science Teacher
and Assistant Principal
Flushing High School
Flushing, New York

Layout, graphics and typesetting: Burmar Technical Corporation, Albertson, NY.

This book is dedicated…

to my great-uncle Dr. Nathan Ronald Brewer, for his abiding love of science — ***Mark Jarrett***

to my wife Joan, my children Todd and Ronald, and my grandchildren Jared and Katie — ***Stuart Zimmer***

to my wife Donna, my children Christian, Carrie, and Jesse, and my grandchildren Aiden and Christian — ***James Killoran***

TABLE OF CONTENTS

UNIT 1: INTRODUCTION

Chapter 1: A Look at the Ohio Science Standards 2
Chapter 2: How to Answer Multiple-Choice Questions 4
Chapter 3: Interpreting Different Types of Data 12
Chapter 4: How to Answer Short and Extended-Response Questions 25

UNIT 2: THE NATURE OF SCIENCE

Chapter 5: Scientific Knowledge and Inquiry 32
Chapter 6: Conducting Scientific Investigations 42
Chapter 7: Science and Technology 55

UNIT 3: EARTH AND SPACE SCIENCES

Chapter 8: The Origins of the Universe 65
Chapter 9: Planet Earth: Systems and Processes 73
Chapter 10: Geologic Time and the Impact of Life on Earth 90

UNIT 4: PHYSICAL SCIENCE

Chapter 11: The Structure of Matter 98
Chapter 12: The Properties of Matter 110
Chapter 13: Force and Motion 124
Chapter 14: The Nature of Energy 133

UNIT 5: LIFE SCIENCES

Chapter 15: Cells and Cellular Processes 149
Chapter 16: Heredity and Genetics 161
Chapter 17: Evolution 170
Chapter 18: Ecology 181

UNIT 6: A FINAL ASSESSMENT

Chapter 19: A Practice OGT in Science 196
Glossary 207
Ohio's Academic Science Standards 212
Index 219

AN INTRODUCTION TO THE OHIO GRADUATION TEST IN SCIENCE

A few years ago, the Ohio General Assembly established the **Ohio Graduation Test** — or OGT. This test was created to make sure Ohio students meet the academic standards they need to graduate from high school. The OGT is given in 10th grade.

Everyone wants to get a high score on the **OGT in Science**. Unfortunately, just wanting a high score on the test is not enough. You really have to work at it! With this book as your guide, you should be better prepared for the test — and maybe even enjoy studying for it.

LEARNING ABOUT THE OGT

Let's start by learning more about the test you will be taking. The **OGT in Science** has four question types. These will test your ability to:

★ ***Recall or identify scientific facts, concepts and relationships.***

★ ***Understand and analyze scientific information.***

For example, summarizing or evaluating data, making inferences from data and observations, describing patterns and trends, and explaining concepts.

★ ***Demonstrate investigative processes of science.***

For example, discussing scientific methods and procedures, and making predictions from texts, graphs, charts and tables.

★ ***Apply concepts and make relevant connections with science.***
For example, using and integrating knowledge and concepts in new situations, and recognizing scientific procedures appropriate to real-world situations.

These question types will be divided among multiple-choice, short-answer and extended-response questions. The test has 32 **multiple-choice questions**, each worth 1 point. **Short-answer questions** require a brief response in which you write a few sentences to answer a question. There are four short-answer questions, each worth 2 points. **Extended-response questions** require a longer response. There are two extended-response questions, each worth 4 points. The test has a total of 48 points.

CHAPTER 1

A LOOK AT THE OHIO SCIENCE STANDARDS

The **OGT in Science** assesses your mastery of six academic standards you have learned in the ninth and tenth grades.

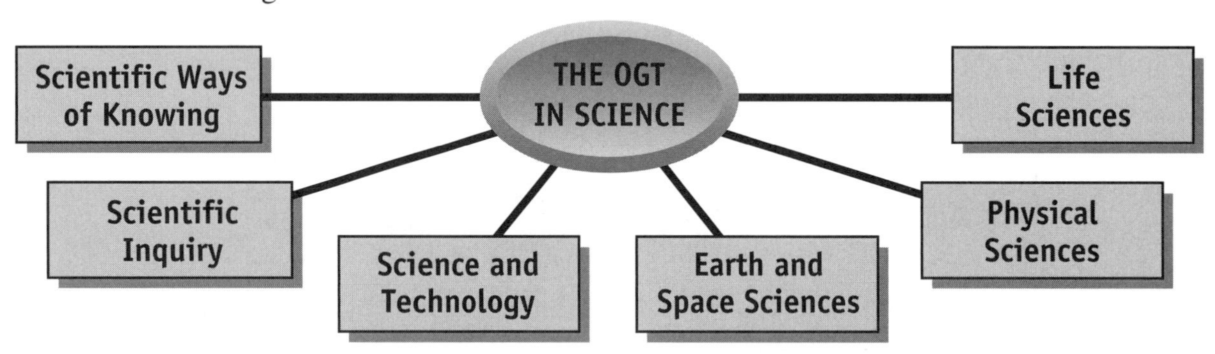

Each standard will have from one to ten benchmarks or particular things you should know about that standard. The questions on the **OGT in Science** will focus on these benchmarks. The distribution of questions and their point values are as follows:

Standard	Multiple-Choice Questions (1 point each)	Short-Answer Questions (2 points each)	Extended-Response Questions (4 points each)	Total Points
Scientific Ways of Knowing; Inquiry; Science and Technology	4	2	1*	12
Earth and Space Science	8–10			12
Physical Sciences	8–10	2	1*	12
Life Sciences	8–10			12
Total Number of Items (38)	32	4	2	48

On any test, 1 of these 3 standards will have an extended-response item.

HOW THIS BOOK CAN HELP YOU

This book is designed to help you perform your best on the **OGT in Science**. You may use this book in the course of the school year as you study the topics that are tested, or you may use this as a final review in the weeks just prior to the test. Here is how the book is organized:

TOOLS FOR MASTERING THE TEST

Unit 1 reviews some of the general skills you need for the test. Chapter 2 explores each of the four types of questions on the test. Chapter 3 explains how to analyze different types of scientific data found on the **OGT in Science** — such as diagrams, line and bar graphs, and tables. Chapter 4 shows you how to answer short-answer and extended-response questions, with sample questions, sample responses and a step-by-step approach.

A REVIEW OF THE BENCHMARKS

The main part of the book has four units surveying each of Ohio's standards and benchmarks.

★ **Unit 2** consists of several chapters dealing with the nature of science. This unit covers ***Science and Technology, Scientific Inquiry,*** and ***Scientific Ways of Knowing***. In this unit, you will learn how scientists think and approach problems. What you learn here will apply to all fields of science.

★ **Unit 3** reviews what you need to know about the ***Earth and Space Sciences***,

★ **Unit 4** covers ***Physical Sciences***.

★ **Unit 5** deals with ***Life Sciences***.

Each content chapter follows a similar pattern of organization. It opens with a "**Major Ideas**" identifying the main ideas you should know on the theme of the chapter. This is followed by a short summary of key ideas and facts. At the close of the chapter, a special "**What You Should Know**" feature again highlights the main points. This is followed by "**Study Cards**." On the back of these cards you should make a diagram, drawing, or picture of your own to help you recall the item. The final section of each chapter presents a model question with a detailed explanation of how to answer it. This is followed by additional OGT-style multiple-choice, short-answer, and extended-response questions.

A FINAL PRACTICE OGT IN SCIENCE

Unit 6 has a practice test to see what you have learned and further prepare you for the actual **OGT in Science**. It has the same distribution of questions and level of difficulty as the real test.

CHAPTER 2

HOW TO ANSWER MULTIPLE-CHOICE QUESTIONS

Most of the questions on the **OGT in Science** are multiple-choice: they ask you to select the best answer from among four choices. There will be four types of multiple-choice questions on the **OGT in Science**:

Recall scientific facts, concepts and relationships	Understand and analyze scientific information	Demonstrate investigative processes of science	Apply concepts and make relevant connections with science

This chapter will help you to recognize and answer each type of question.

RECALLING INFORMATION

Many questions on the **OGT in Science** will assess your knowledge of important scientific facts, concepts and relationships. Some questions on the test will simply ask you to recall or identify a scientific fact, concept, or relationship.

Examine the following sample question, which tests your ability to identify an important concept:

1 Green plants transform light energy into chemical energy through the process of

A. fermentation
B. photosynthesis
C. respiration
D. excretion

UNLOCKING THE ANSWER

To answer questions asking you to recall information, consider using the three-step "**E-R-A Approach**" — ***Examine***, ***Recall***, and ***Apply***.

Step 1: EXAMINE the Question
Carefully read the question. Make sure you understand all the information it includes. Focus on what it specifically asks for. Also look at the four answer choices. Eliminate any choices that you know are wrong.

Question 1 asks about the process that transforms light energy into chemical energy in green plants. Notice that the question describes the process, while the answer choices give different names for this process. Your task is to select the correct name.

Step 2: RECALL What You Know
Now identify the subject that the question asks about. Take a moment to think about what you know about that subject. Mentally review the most important *concepts*, *facts*, and *relationships* that you can remember.

This question asks about plant life. What do you remember about this subject? You might recall that plants use sunlight to turn carbon dioxide and water into glucose and other carbohydrates. Energy from sunlight is stored in the chemical bonds of these organic compounds.

Step 3: APPLY What You Know to Answer the Question
Apply the information that you remember to select the correct answer.

To answer this question, you must recall that **photosynthesis** *is the name given to the process of turning sunlight into chemical energy. The correct answer is therefore choice B.*

As you can see, it will be quite important to remember major facts, concepts, and relationships to answer this kind of question correctly. This book includes several tools to help you answer this question type:

- ★ **Content Units.** Each content unit summarizes the most important facts, concepts, and relationships in an important field of science.
- ★ **Major Ideas.** Appearing at the start of each chapter, this feature highlights the most important information about each benchmark that you need to know.
- ★ **What You Should Know**. A separate feature at the end of each content chapter reviews the most important facts, concepts, and relationships in that chapter.

★ **Study Cards.** You will also find *Study Cards* at the end of each chapter. Use these cards — alone or with friends — to learn and review the most important facts, concepts and relationships in each area of science. It will help if you make a drawing or diagram of your own on the back of each card to illustrate important concepts and facts.

★ **Practice Questions.** Practice questions at the end of each chapter will help you see if you can recall the most essential information in the chapter.

ANALYZING SCIENTIFIC INFORMATION

To ***analyze*** means to break something up into its various parts in order to understand it better. Some questions on the **OGT in Science** will ask you to analyze scientific information and to show that you understand it. Very often, this kind of question will present you with scientific data or observations and ask you questions about them. These questions may ask you to:

- Organize, summarize or evaluate results from an experiment
- Make estimates based on data
- Choose which model best represents a sample of data
- Draw conclusions from information
- Describe patterns or relationships in observations or data, such as cause-and-effect relationships

Let's look at a sample question asking you to analyze scientific information.

2 **As the car rolls down the hill, its kinetic energy**

A. increases while its gravitational potential energy decreases.

B. remains the same while its gravitational potential energy increases.

C. increases while its gravitational potential energy remains the same.

D. decreases while its gravitational potential energy increases.

UNLOCKING THE ANSWER

To answer questions asking you to analyze scientific information, again use the **"E-R-A Approach"**:

Step 1: EXAMINE the Question
Read the question carefully. Examine any data or observations it may include. Usually, this data will be presented in the form of a diagram, graph, or table. Make sure you understand what the data shows and what information the question asks for about the meaning of key terms. Don't forget to check the answer choices.

Step 2: RECALL What You Know
Next, identify the subject of the question. Take a moment to think about any important facts, concepts and relationships you can recall about the subject of the question.

You should recognize that this diagram shows energy and motion. The answer choices focus on kinetic and potential energy. You should recall that kinetic energy increases as the speed of a moving object increases ($KE = \frac{1}{2}mv^2$). You should also recall that a body's gravitational potential energy (stored energy based on its position) ***decreases*** *as it rolls downhill.*

Step 3: APPLY What You Know to Answer the Question
Apply the information that you remember to select the correct answer.

In this case, you have to select the choice that correctly identifies what happens to both the kinetic energy and the gravitational potential energy of the car. As the car rolls downhill, its kinetic energy increases, but its gravitational potential energy decreases. The correct answer choice is A.

EXPLAINING SCIENTIFIC INVESTIGATION

Some questions on the **OGT in Science** will test your ability to think like a scientist. The next page shows some of the things you may be asked to do:

Make observations	Describe the procedures used in an experiment	Make scientific measurements	Make predictions or develop questions based on a scientific experiment

For example, look at the following two questions. Each one deals with a different aspect of scientific investigation:

A student investigated the effectiveness of four different mouthwashes in destroying bacteria. He placed the nutrient agar in four petri dishes containing bacteria. Each of four paper disks, 1 centimeter in diameter, was soaked in a different mouthwash sample and placed on a different agar surface. Each petri dish was placed in an incubator at a temperature of 37°C for a 24-hour period. The diagram below represents the sequence of events in this investigation. The shaded areas in the petri dishes represent regions of bacterial growth.

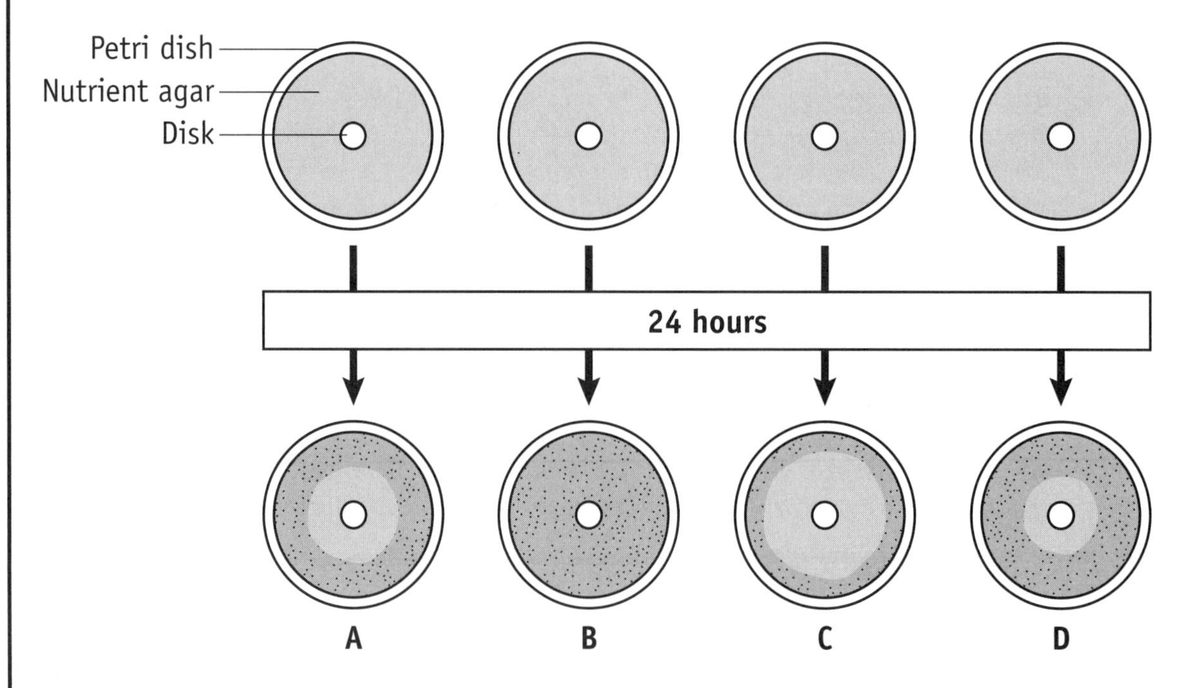

3 Based on the diagram, which petri dish contains the most effective mouthwash?

A. dish A

B. dish B

C. dish C

D. dish D

- Examine the Question
- Recall What You Know
- Apply What You Know

4 An object was placed in a graduated cylinder containing 100 milliliters of water. The diagram to the right illustrates the new level of water.

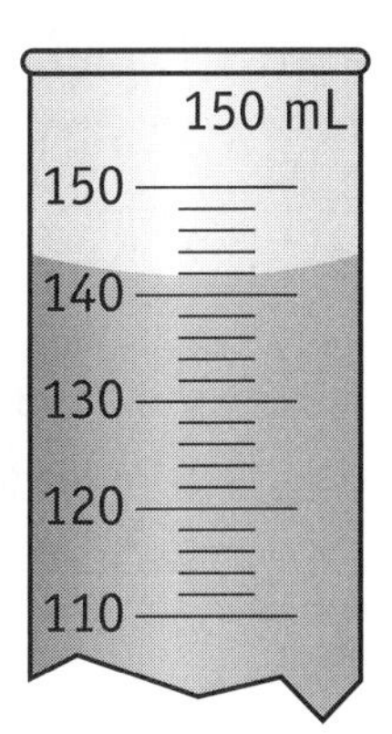

What is the volume of the object?

A. 42 mL
B. 44 mL
C. 142 mL
D. 144 mL

UNLOCKING THE ANSWER

To answer questions asking you to analyze scientific information, you should again use the "**E-R-A Approach**":

Step 1: EXAMINE the Question
Read the question carefully, examining any data it may include. Before you try to answer the question, it is essential that you understand what information the data is showing.

In question 3, you are asked to examine a group of Petri dishes and draw a conclusion, while question 4 asks you to examine a graduated cylinder and make a measurement.

Step 2: RECALL What You Know
Now think about the subject of the question. Does the question ask you to make an observation, a measurement or a prediction? Or does it ask you to describe a scientific procedure? Take a moment to think about what you know about scientific investigations.

For question 3 you have to think about what results you would expect from the most effective mouthwash. For question 4, you have to think about how to determine the volume of an object, using a graduated cylinder. Remember to read the volume from the bottom of the concave surface (meniscus).

Step 3: APPLY What You Know to Answer the Question
Apply the information that you remember to select the correct answer.

Both questions ask you to apply your basic knowledge of scientific investigation. The answer to question 3 is choice C: the diagram shows that this mouthwash allowed the least bacterial growth since its shaded area is smaller than the others. What is the answer to question 4?

In Chapter 6 of this book, you will learn more about methods of scientific investigation and laboratory procedures.

APPLYING SCIENTIFIC CONCEPTS TO "REAL WORLD" SITUATIONS

Some questions on the **OGT in Science** will ask you to apply scientific concepts to "real-world" situations. These questions could ask you to:

- Apply your scientific knowledge to new situations
- Use scientific concepts to solve problems
- Determine which scientific procedures to use in an investigation

QUESTIONS DEALING WITH "REAL WORLD" SITUATIONS

Look at the two sample questions below. Each question asks you to make connections between science and the "real world."

5 A world health organization is concerned about the effects of increasing amounts of carbon dioxide in Earth's atmosphere. What step can they take to reduce these amounts?

A. discourage the use of chemical pesticides
B. encourage the reforestation of tropical rainforests
C. encourage the planting of farm crops in place of forests
D. discourage the use of nuclear energy as a source of power

- ♦ Examine the Question
- ♦ Recall What You Know
- ♦ Apply What You Know

6 Many communities have attempted to control the size of mosquito populations to prevent the spread of such diseases as malaria and West Nile Virus. Which control method is likely to cause the *least* ecological damage?

A. draining the swamps where mosquitoes breed
B. spraying swamps with chemical pesticides to kill mosquitoes
C. spraying oil over swamps to suffocate mosquito larvae
D. increasing populations of native fish that feed on mosquito larvae

UNLOCKING THE ANSWER

To answer questions asking you to apply scientific concepts to "real-world" situations, again apply the "**E-R-A Approach**":

Step 1: EXAMINE the Question
As always, read the question carefully, examining any data it may include. Before you try to answer the question, it is essential that you understand what information the question presents. Then focus on what the question asks for.

In this case, both questions ask about steps that can be taken to protect the environment.

Step 2: RECALL What You Know
This kind of question asks you to apply your scientific knowledge to "real-world" situations. You need to identify which scientific facts, concepts, and relationships best apply to the problem or situation. Take a moment to review the appropriate facts, concepts, and relationships in your mind.

To answer question 5, you might recall from Life Sciences that plants absorb carbon dioxide in the process of photosynthesis.

To answer question 6, you should recall what you learned about the interdependence of life in ecological systems. Introducing changes in one part of an ecosystem may have effects throughout the ecosystem.

Step 3: APPLY What You Know to Answer the Question
Now you are ready to apply what you recalled to the question. Think about how the scientific facts, concepts, or relationships you recalled can be used to resolve the problem posed by the question.

In question 5, which answer choice helps reduce carbon dioxide in Earth's atmosphere? Because trees absorb more carbon dioxide than farm crops do, many environmentalists are troubled by the destruction of the world's tropical rainforests. Reforesting rainforest would help to absorb more carbon dioxide. Therefore, the correct answer choice would be B.

In question 6, you must find the answer choice that best controls the size of the mosquito population without upsetting the ecological balance. Draining the swamp or using chemicals would have serious ecological consequences. Which answer choice do you think is best?

CHAPTER 3

INTERPRETING DIFFERENT TYPES OF DATA

In the last chapter, you learned about four types of questions on the **OGT in Science**. Many of these questions will provide observations or data that you have to interpret. In this chapter, you will learn how to interpret various ways that data can be organized and presented. These formats include the following:

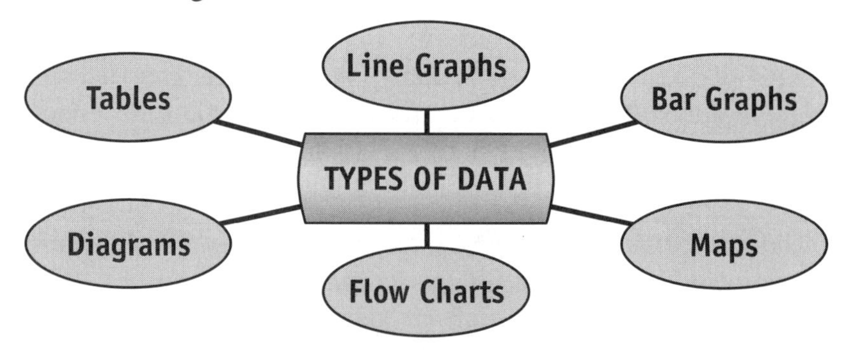

Remember to use the **"E-R-A Approach"** — *Examine*, *Recall*, and *Apply* — when answering any question that includes data:

UNLOCKING THE ANSWER

Step 1: EXAMINE the Question
Carefully read the question and examine the data. This chapter will help you to understand different ways that data can be presented. In this first step, you should focus on what the question asks for.

Step 2: RECALL What You Know
Next, identify the subject that the question asks about. Take a moment to think about what you know about that subject. Mentally review the most important concepts, facts, and relationships that you can remember.

Step 3: APPLY What You Know
Apply the information you recalled from your prior knowledge of science to answer the question.

TABLES

A **table** is used to organize information, especially quantitative (*numerical*) data. Tables list information in columns and rows. Each column is labeled, and sometimes rows are also labeled. To interpret a table, you should pay close attention to the headings for the columns and rows, and to any units of measurement that are used.

Variables. A scientific table often shows the relationship between two things. Because both items can change, or *vary*, they are known as **variables**. Scientists usually have control over the first variable — known as the **independent variable**. They change this first variable to see what effects this has on the second variable — known as the **dependent variable**.

At the same time, scientists try to hold all other factors constant. This allows them to really see what effect changes in the independent variable have on the dependent variable. Let's see how this applies by examining the following table and information:

A group of scientists has been studying the amount of oxygen that can be dissolved in water at different temperatures. They have gradually heated the water and measured how much oxygen could be dissolved in it at 5° Celsius intervals.

AMOUNT OF OXYGEN THAT CAN BE DISSOLVED IN WATER

Water Temperature (C°)	Oxygen Content (*parts per million*)
10	11.29
15	10.10
20	9.11
25	8.27
30	7.56

In this experiment, scientists measured the amount of oxygen that could be dissolved in fresh water at various temperatures. They recorded their results in the table.

- ★ The first column shows the **independent variable** — the temperature of the water in degrees Celsius (C°).
- ★ The second column shows the **dependent variable** — the amount of oxygen that could be dissolved, in parts per million.

"Parts per million" means that in a given amount of water, one-millionth of it is dissolved oxygen. On the next page, we will see how a question might be asked about this data.

1 According to the table, how much oxygen can be dissolved in water at 20°C?

A. 9.11 parts per million
B. 8.27 cubic centimeters
C. 9.11 millimeters
D. 8.27 mL

- Examine the Question
- Recall What You Know
- Apply What You Know

This question tests your understanding of the table and its units of measurement. At 20°C, the amount shown on the table in the second column is "9.11". According to the heading at the top of the second column, the number 9.11 refers to "parts per million." This means that in one million parts of water at 20°C, no more than 9.11 parts per million could be dissolved oxygen.

2 According to the data in the table, what happens to the amount of oxygen that can be dissolved in water as its temperature rises?

A. the amount increases
B. the amount decreases
C. the amount stays the same
D. the amount decreases, then increases

This question asks you to ***analyze*** the information in the table. To answer this question correctly, study the table and try to determine the relationship between the independent variable (*temperature*) and the dependent variable (*oxygen content*). The clue to answering this question correctly is to find out what happens to the temperature as you move down the table. What happens to the oxygen content as the temperature rises?

3 How much oxygen can be dissolved in water at a temperature of 23°C?

A. 9.57 parts per million
B. 14.24 parts per million
C. 8.59 parts per million
D. 7.50 parts per million

This question asks you to predict the oxygen content for a temperature not found directly on the table. To answer this question correctly, you must use information already provided in the table to infer something. Because 23° is between 20° and 25°, you know the oxygen content must be less than at 20°C but greater than at 25°C. Which of the answer choices meets these conditions?

LINE GRAPHS

A *line graph* shows a series of points on a grid connected by a line. The title tells what the graph shows. Each point on the grid represents a specific quantity. The purpose of a line graph is to show how two or more variables are related.

Usually, a line graph shows how changes in one variable will lead to changes in the other variable. The **X axis** usually represents the *independent variable*, and the **Y axis** represents the *dependent variable*. The X axis is the bottom line of the graph. As you move to the right of the graph, the value of X on this line increases. The left side of the graph is the Y axis. As you move up this line, the value of Y increases.

The graph below shows information about ocean temperatures. Here, scientists have measured the temperature of the ocean at different depths. Which is the independent variable? Which is the dependent variable? As you can see from this example, a line graph often shows patterns better than a table does.

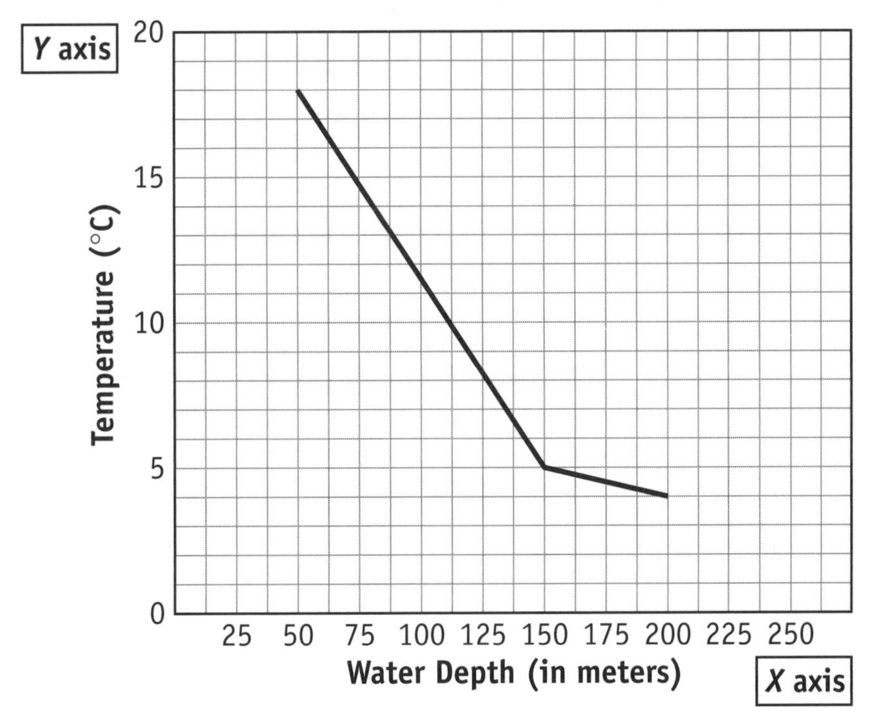

4 Based on the graph, what happens to the water temperature as the ocean depth increases?

A. The temperature of the ocean water remains constant.

B. The temperature of the ocean water decreases.

C. The temperature of the ocean water increases.

D. The temperature of the ocean water increases, then decreases.

- Examine the Question
- Recall What You Know
- Apply What You Know

As you have just read, each **axis** represents a different variable. The *horizontal* (*or X*) axis shows the depth of the water in meters. The *vertical* (*or Y*) axis shows the temperature of the ocean at that depth. Each point on the graph represents the temperature (*the dependent variable*) at a particular water depth (*the independent variable*). Now that you understand this, can you determine what happens to the temperature as the depth of the ocean's water increases?

APPLYING WHAT YOU HAVE LEARNED

Since you now know how to read a line graph, let's see if you can also make one. Review the table on page 13 and turn this information into a line graph.

Very often, a line graph will show how something changes over time. In this case, time is the "independent variable." Sometimes there may even be several lines to show how different variables relate to each other over time. For example, look at the graph below. It shows the deer population in a park area between 1900 and 1945. The lines indicate how two variables changed with the passage of time: the actual number of deer and the "carrying capacity" of the park to support the deer population. The "carrying capacity" of an area refers to how many of a particular kind of animal the area can support.

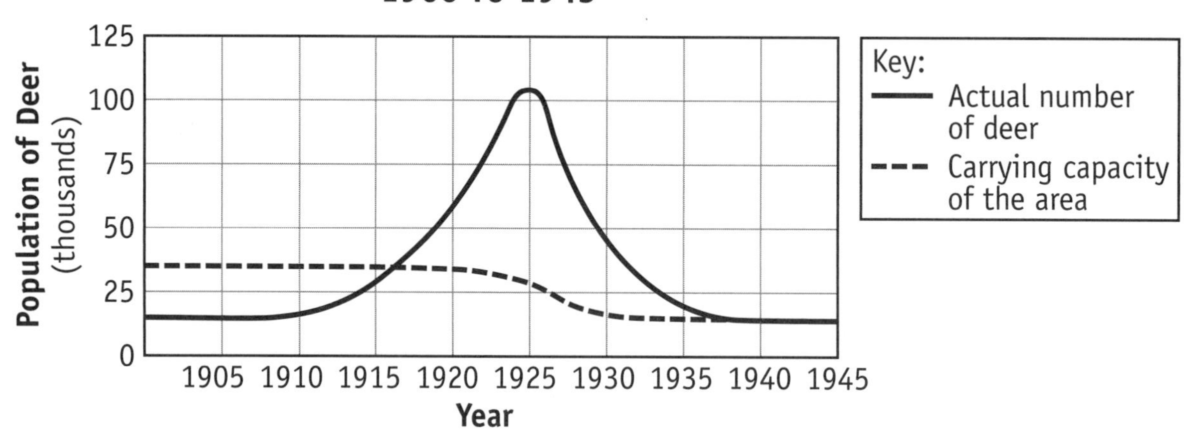

5 **Which statement most likely explains why the carrying capacity of Jones Park to support deer decreased after 1925?**

A. The deer population in the park decreased in 1926.
B. The number of predators increased between 1915 and 1925.
C. The deer population had grown so large that it ate much of the park's vegetation.
D. An unusually cold winter occurred in the park in 1918.

- Examine the Question
- Recall What You Know
- Apply What You Know

Here, the **key** tells us that the solid line represents the actual number of deer in Jones Park, while the broken line represents the carrying capacity — the number of deer that Jones Park can support. The carrying capacity of the area remained about the same between 1900 and 1920. However, by 1925 the park had 100,000 deer. This greater number of deer ate much of the vegetation. Once most of the vegetation was gone, the park could no longer feed so many deer. For this reason, the correct answer choice is C.

BAR GRAPHS

A **bar graph** is made up of parallel bars of different lengths. It is often used to compare something over time. Each bar represents a quantity. Either each bar is labeled or a key explains what each bar represents. The Y axis often indicates what quantities the bars show.

Use the bar graph below to answer the question.

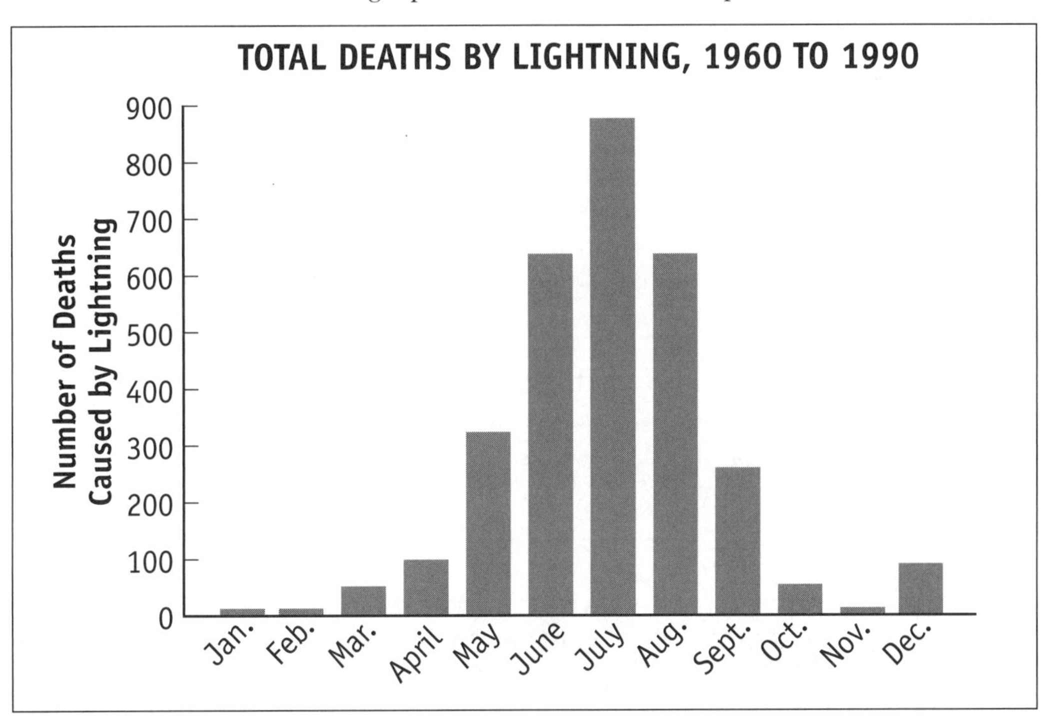

6 What conclusion can be drawn from the information in the bar graph?

A. Most deaths from lightning occur along the eastern coastline.

B. The average number of deaths each year is about 900.

C. The greatest number of deaths generally occur in summer months.

D. Most deaths from lightning occur as a result of hurricanes.

- Examine the Question
- Recall What You Know
- Apply What You Know

This question tests your ability to draw conclusions from a bar graph. Here, each bar shows the number of deaths from lightning in a particular month over a 30-year period. Thus, 900 people died from lightning strikes in July between 1960 and 1990. To answer the question correctly, you should use the information in the bar graph to test each answer choice. For example, the graph does not tell whether most deaths from lightning occurred on the eastern coastline. Answer choice A is therefore wrong.

DIAGRAMS

A **diagram** is a simplified picture that shows how several things are related or that shows the different parts of a single thing. Arrows are often used to indicate important relationships. The purpose of a diagram is to help the reader visualize how something works. For example, look at the diagram below. It represents the direction of Earth's rotation if one were looking down from outer space just above the North Pole.

7 Based on the diagram, the time at point X is closest to

A. 9 A.M.

B. 12 noon

C. 9 P.M.

D. 12 midnight

- Examine the Question
- Recall What You Know
- Apply What You Know

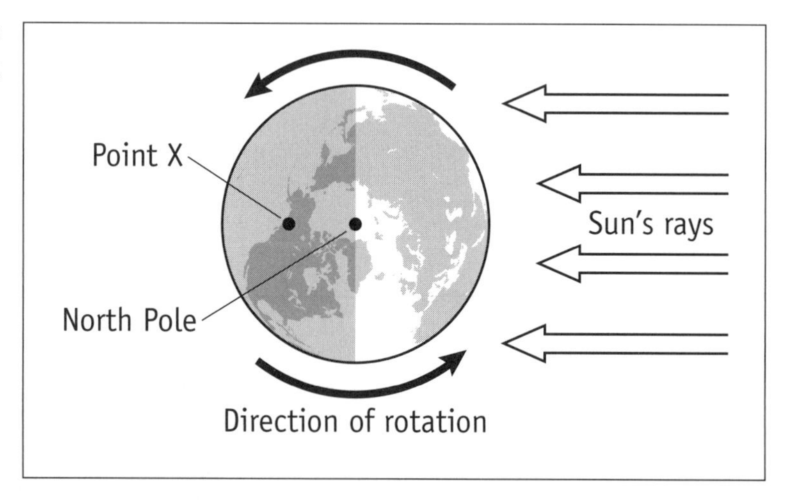

To answer this question, you have to understand what the diagram shows. Curved arrows indicate the direction Earth rotates, while straight arrows indicate the direction of the sun's rays. The shaded area indicates night. You can quickly eliminate choices A and B, because Point X is clearly nighttime. Choice D is a better choice than choice C, because Point X is exactly in the middle of the shaded area — exactly opposite the sun's rays.

ANSWERING CLUSTER QUESTIONS

Sometimes you may come across a **cluster** (*group*) of questions on the **OGT in Science**. These questions may be about a single piece of data, or about several related graphs, tables, or written information. When you come across a cluster of questions, you should still use the **E-R-A Approach**. However, there are special things to keep in mind approaching cluster questions or multiple pieces of data.

STEP 1: ANALYZE THE QUESTION.

★ **Look at the Background Information.** First, carefully read **all** the information provided as background to the question before you attempt to answer any of the questions. If the question is about an experiment, be sure you understand what the experimenters were trying to find out and what they did.

★ **Read and Take Notes:** When reading introductory information, use the skills you use in other subjects to understand it: underline key words and take brief notes in the margin. If you are reading a very complex passage, you might want to briefly take separate notes or outline what you have read.

★ **Examine all Graphs and Diagrams.** Pay special attention to any graphs. Make sure you understand what the X and Y axis on each graph represents. Look at the units of measurement being used in the data.

★ **Make Connections.** Make connections by asking yourself: What is the data mostly about? How are these different charts, graphs, tables and text related?

STEP 2: RECALL WHAT YOU KNOW.

Now you are ready to think about those scientific concepts, relationships, and facts you know that relate to the subject of the question.

STEP 3: APPLY WHAT YOU KNOW.

Finally, apply your understanding of the information provided in the cluster and your knowledge of science to answer each question.

★ **Focus on the Important Things.** Usually questions will test important scientific principles or ideas and **not** an insignificant piece of information.

★ **Refer Back Often.** Be sure to refer back to the introductory information, the pieces of data and your own notes as you answer each question in the cluster.

★ **Put on Your Thinking Cap.** Remember, some answers to cluster questions will be stated **directly** in the data — the graph, table or text. Other questions will ask you to use the information supplied to make an **inference**, **deduction** or **conclusion** that may not be directly stated. In answering these questions, rely on both your knowledge of science and your own common sense.

The world's ocean floor were once unknown to scientists. In recent years, new technologies have allowed scientists to explore these regions. Small, unmanned submersible vehicles travel as deep 4 km to explore underwater volcanoes, mountain ranges, trenches, and plains. Computers use data from submersibles to generate pictures of the ocean floor. Scientists also use information from seismic waves — waves of energy sent out by earthquakes — to map Earth's interior. The diagram below is based on information that scientists have learned about the ocean floor in the middle of the Atlantic Ocean.

The diagram below shows the parts of a cross-section of a rift valley in the center of a mid-ocean ridge. It is followed by a cluster of questions.

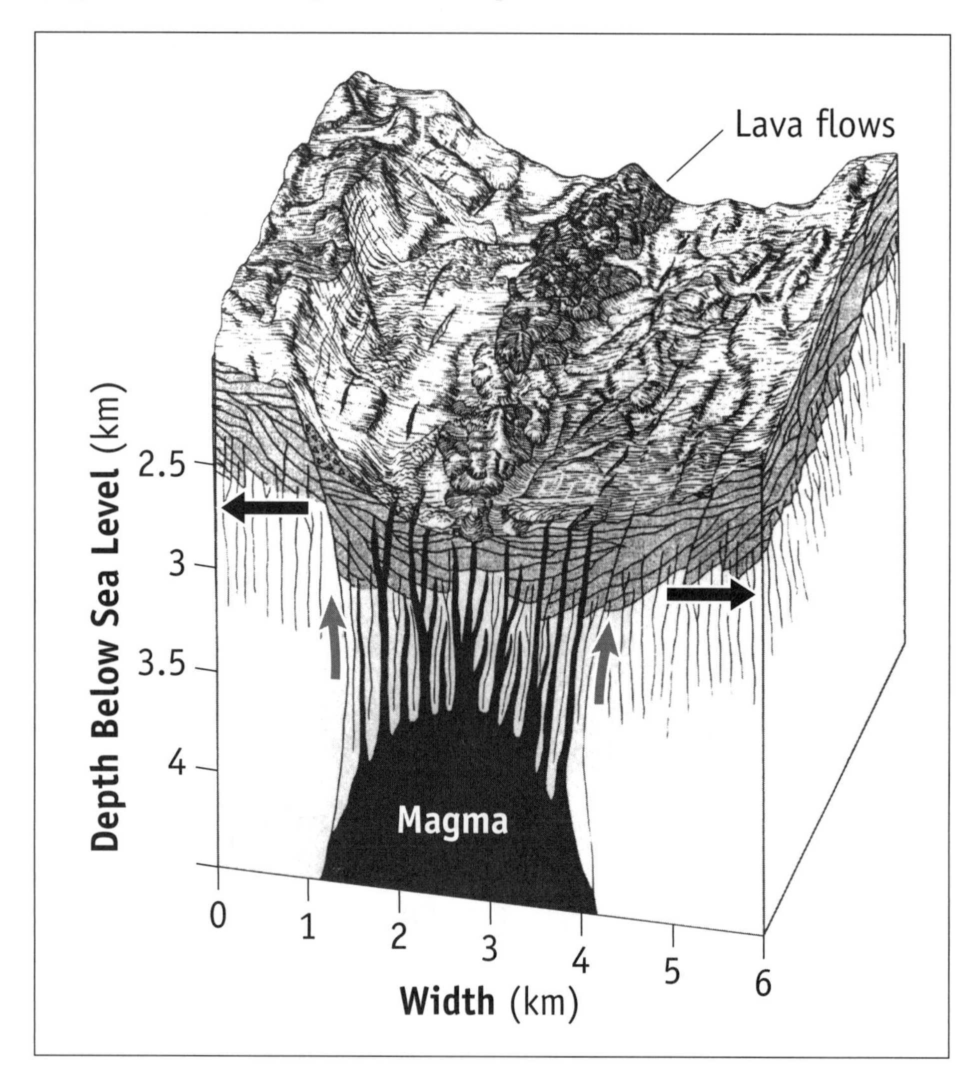

As you study this diagram, you should notice that it is far more detailed than the diagram on page 18. It looks like a realistic picture of a slice of Earth. The vertical lines in the diagram represent faults and fractures in the ocean floor bedrock. A line identifies a ridge at the top of the bedrock, created by lava flows. The scale on the left of the diagram shows the depth of the section. The black material at the bottom of the cross-section is identified as **magma** — molten rock below Earth's crust. Arrows show the movement of magma and the plates.

Now that you understand what this cross-section shows, answer the following questions that could be asked about this diagram.

8 According to the diagram, how far have the crustal plates dropped in the center of the cross-section, from their original depth of 2.5 km?

A. 3 km
B. 1 km
C. 0.5 km
D. 0 km

◆ Examine the Question
◆ Recall What You Know
◆ Apply What You Know

Question 8 tests your ability to interpret the diagram. The gray area in the center of the diagram has dropped from 2.5 to a level of about 3.0 km. Therefore, the correct answer choice is C.

9 Which type of tectonic plate boundary is shown in this diagram?

A. divergent
B. universal
C. convergent
D. transform

To answer question 9, you have to recall your knowledge of Earth Science to correctly identify the type of tectonic plate boundary. Since the plates are moving in opposite directions, they are separating or **divergent**. Divergent plate boundaries are where the plates move apart from each other, creating a gap that is filled by magma. Therefore, the correct answer choice here would be A.

10 What will be the probable result of the continuation of the geologic processes indicated at this location?

A. Earth's magnetic field will reverse direction.
B. Continental crust will be forced downward.
C. Earth's circumference will increase.
D. New oceanic crust will form.

Question 10 asks you to predict the effects of the forces shown in the diagram. If you carefully examine the diagram, you will see two arrows indicating that the hot, molten magma is pushing upward. The direction of the lines on the ocean floor bedrock show it is being pushed apart in opposite directions. Which answer correctly identifies the result of this process?

FLOW CHARTS

A **flow chart** is a special type of diagram. It shows a series of steps in a process. Each step is placed in some geometric shape or represented by a picture. The shapes or pictures are connected by arrows or lines to indicate the sequence of steps in the process. Sometimes the process may proceed in more than one way. Arrows will indicate when a step may lead to different choices or outcomes.

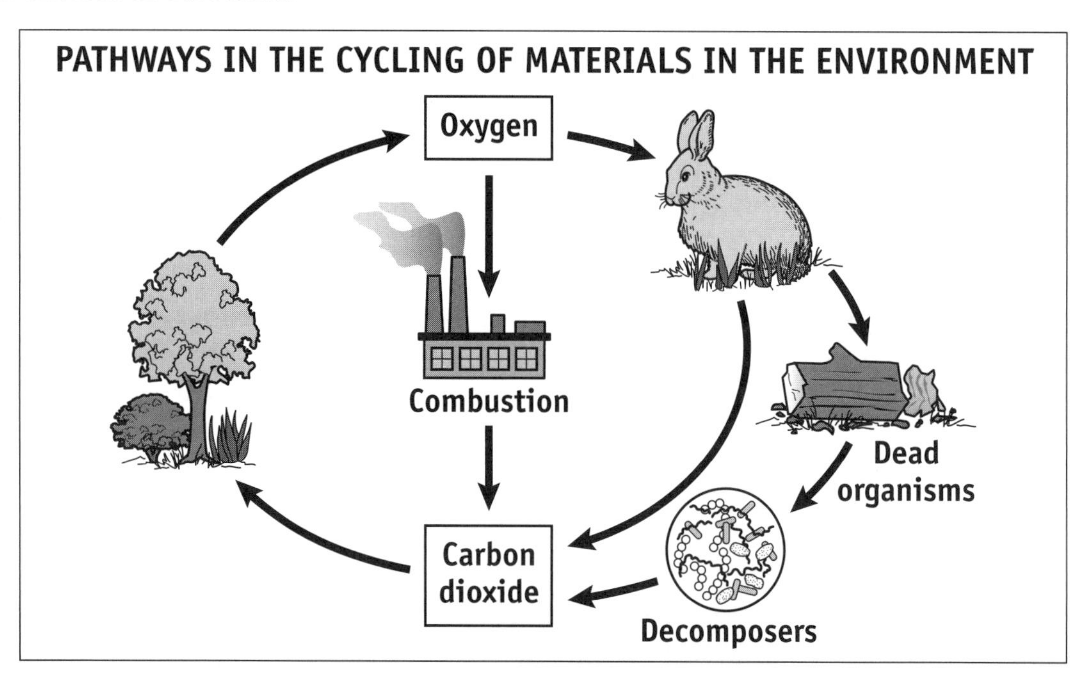

11 Based on the flow chart above, which of the following is a source of oxygen in the atmosphere?

A. combustion
B. trees
C. decomposers
D. animals

- Examine the Question
- Recall What You Know
- Apply What You Know

To answer this question, you have to understand how a flow chart works. The arrows indicate how organisms produce and consume two gases — oxygen and carbon dioxide. Using the flow chart, notice that:

- Plants, such as trees, consume carbon dioxide and produce oxygen.
- Animals and combustion processes consume oxygen and produce carbon dioxide.
- Decomposers produce carbon dioxide when they break down dead organisms.

Based on these facts, which answer is correct?

MAPS

A **map** is a diagram representing a place or area. It shows where objects are located. Maps may be used to show the location of the stars or planets, or to depict characteristics of Earth's surface. The **legend** of the map explains the meaning of symbols used on the map. The **direction indicator** or **compass rose** shows directions on the map. The **scale** tells what distances on the map represent in the actual area.

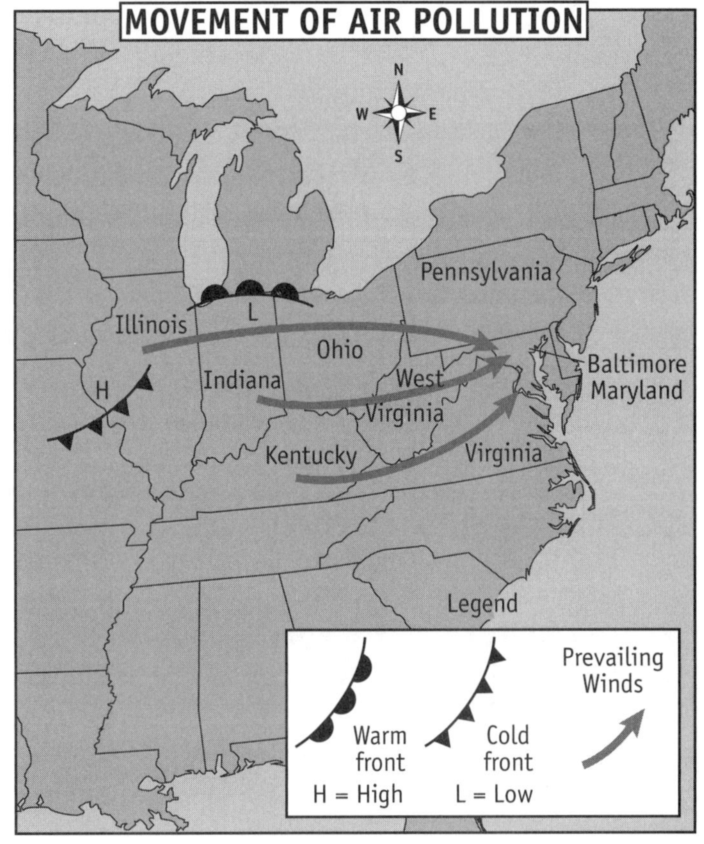

Source: Environmental Protection Agency

12 Which conclusion can be drawn from the map?

A. Illinois produces more air pollution than other states.
B. The air pollution problem in Baltimore is increased by the addition of pollution from other areas.
C. There is no air pollution in any southern states.
D. Air pollution problems in Virginia clear up quickly as the air moves toward the sea.

13 Based on the information in the map, in the next few days southern Illinois is likely to have

A. warmer weather
B. heavy precipitation
C. colder weather
D. a hurricane

To answer question 12, you have to study the information on the map. The arrows show that winds push air pollution from the Midwest across the eastern United States. From the information, we cannot conclude that choices A, C or D are correct. We can determine that these winds appear to converge on Baltimore. The map also shows a cold front moving from the northwest and a warm front moving from Ohio northwards. Which of these two fronts is about to hit southern Illinois?

TESTING YOUR UNDERSTANDING

Use the information below to answer the following question.

For several months, a small Ohio community was heavily infested with mosquitoes. They began spraying weekly with insecticide. Daily counts providing information on mosquito population size are represented in the graph below.

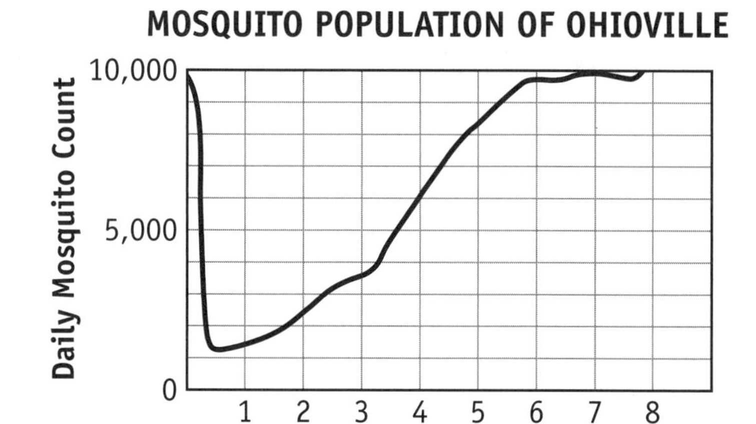

1 What is the likely reason for the decreased effectiveness of the insecticide?

A. It caused mutations in the mosquitoes, which resulted in immunity.

B. It was only sprayed once.

C. Mosquitoes resistant to the insecticide lived and produced offspring.

D. The insecticide chemically reacted with the DNA of the mosquitoes.

2 In the food web to the right, arrows indicate where a food source is consumed. Which organism is classified as both a primary and a secondary consumer?

A. hawk

B. raccoon

C. snake

D. mouse

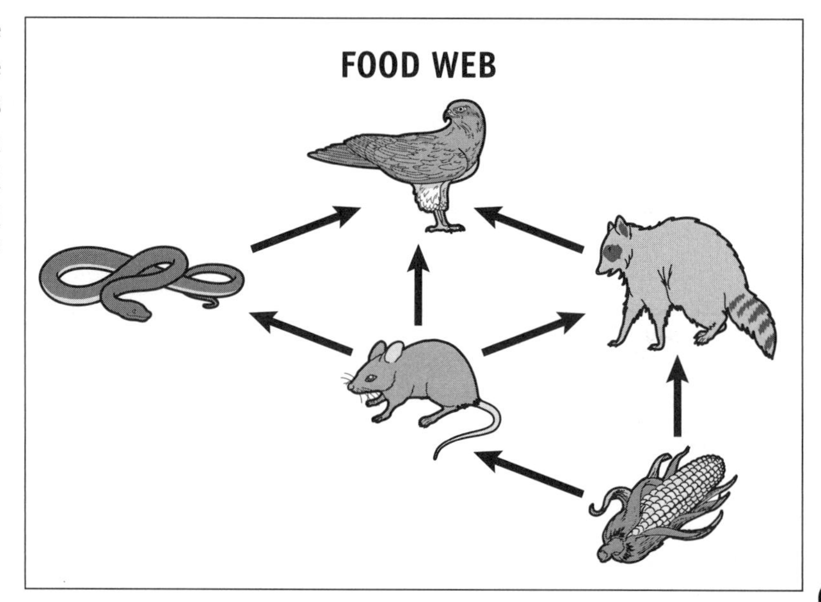

- Examine the Question
- Recall What You Know
- Apply What You Know

Hint: *The direction of arrows in a food web usually shows the flow of nutrients and energy. For example, corn is eaten by a mouse, so its energy and nutrients are absorbed by the mouse. The arrow therefore points from the corn to the mouse.*

CHAPTER 4

HOW TO ANSWER SHORT AND EXTENDED-RESPONSE QUESTIONS

Some questions on the **OGT in Science** will ask you to write an answer in your own words. These questions will typically require more time to answer than a multiple-choice question, and they count for more points.

SHORT-ANSWER QUESTIONS

There will be four short-answer questions on the **OGT in Science**. A short-answer question requires that you give two facts or pieces of information. Each question will be worth two points. You can recognize a short answer question because it will be followed by: "(*2 points*)." Separate short-answer questions will test the following standards:

Two of the following three:

- Earth and Space Sciences
- Life Sciences
- Physical Sciences

Two of the following:

- Scientific Inquiry
- Scientific Ways of Knowing
- Science and Technology

EXTENDED-RESPONSE QUESTIONS

There will be at least two extended-response questions on the OGT. An extended-response question will require you to give four facts or pieces of information. Each of these questions will be worth four points and will be followed by: "(*4 points*)." They will test ***Scientific Inquiry, Scientific Ways of Knowing*** or ***Science and Technology,*** and at least **one** other standard — either ***Earth and Space Sciences, Life Sciences,*** or ***Physical Sciences***.

RESPONDING TO A SHORT-ANSWER OR EXTENDED-RESPONSE QUESTION

There are many ways to approach answering short-answer or extended-response questions.

One of the most common ways is to use three main steps:

STEP 1: ANALYZE AND PLAN

When answering either a short-answer or extended-response question, you must first look very carefully at the directions of the question. The exact instructions for what you are supposed to do will usually be found in the **"performance verbs"** of the question. These performance verbs ask you to present information in a certain way.

SOME OF THE MOST COMMON OGT "PERFORMANCE VERBS"

Analyze	To think about the parts of something to figure out what it is like as a whole.
Compare	To identify similarities and differences of two or more things.
Describe	To give characteristics of something, including ideas or events. A description can also tell how something changes over time.
Explain	To make clear or give the reasons for something. To explain how something happens; tell the way in which it occurs. To explain why something happens; give the reasons why it occurs.
Evaluate	To determine the value or worth of something.
Identify	To name something, or tell what something is.
Predict	To use what is known to make a statement about what will happen in the future.
Summarize	To restate information in a shorter form by stating only what is most important.
Support	To give facts, examples, and other evidence to back a conclusion, argument or point of view.

After you have studied the "performance verb" in the question, you should next identify **all** the parts of the question. Then take a few moments to think about what you know about the subject of the question. Now you are ready to plan your answer. This can be done by simply jotting down notes with details you think will be helpful. It often helps to plan your answer with an **answer box**, showing the different parts of the question. Fill in the answer box with your ideas or simply check off each part of the box as you complete that section of your response. The answer box serves as a checklist, ensuring that you answer each part of the question.

★ **Short-answer questions** will have **two** parts. For example, a short-answer question might present *a laboratory experiment and ask you to **identify** two important measurements that students need to make,* or it might ask you to ***identify*** *one piece of evidence supporting the Big Bang Theory and **explain how** this evidence supports the theory*. Here is what an answer box might look like for this second question:

Piece of evidence	Your Response
How it supports the Big Bang Theory	Your Response

★ **Extended-response questions** will usually have **four** parts. For example, an extended-response question might present you with a description of an ecosystem and ask you to ***identify*** two environmental changes and ***describe*** *one effect of each change on that system*. Or you might be asked to ***identify*** *two effects of plate tectonics and to **explain** how each effect is caused*. Here is what an answer box might look like for this second question:

FIRST EFFECT		SECOND EFFECT	
Identify	Your Response	***Identify***	Your Response
Explain	Your Response	***Explain***	Your Response

STEP 2: WRITE YOUR ANSWER

The next step in responding to a short-answer or extended-response question is to ***write*** your answer. You might use the notes you put in your answer box to write your answer. For the **OGT in Science**, it will not always be necessary to write out your answer in full paragraphs. It might be enough to copy the main parts of your answer box onto the Answer Document. Whichever method you select, make sure it is clear how this information answers the question.

If you decide to write out a full answer, it may help to **"echo" the question**. To echo the question, repeat it in the form of a positive statement. For the example above, you could begin: *"Two effects of plate tectonics are the spreading sea floor and earthquakes along plate boundaries."* Then turn each point of your notes or answer box into one or more complete sentences. Check off sections of your answer box each time you complete that part of your answer.

STEP 3: REVIEW AND REVISE YOUR ANSWER

The first person to read your writing should be **YOU** — ***not*** the person scoring your answer. Once you have finished writing, read over your answer. Make sure you provided all the information required by the question. As you review what you have written, ask yourself:

Did I complete **all of the parts** in the question?	Did I provide enough details, examples, and reasons to **support** my answer?

HOW YOUR ANSWER WILL BE SCORED

To see how your answer will be scored, let's look at a model OGT-style question:

Identify two important scientific theories and provide one example of evidence supporting each theory. Respond in the space provided in your Answer Document. (4 points).

Test scores will use a **rubric**, or scoring guide, to score student responses on the **OGT in Science**. The rubric tells a scorer what information an extended-response should include to receive a score of **0, 1, 2, 3, or 4 points**. Below are sample response data for the question above. The rubric appears on the next page.

SAMPLE RESPONSE (EXTENDED-RESPONSE)

The response may include any two of the following:

- **Heliocentric Theory.** The theory that Earth moves around the Sun was supported by the observations of astronomers using telescopes and charting the movements of the planets.
- **Newtonian Mechanics.** Newton's theory of gravity and laws of motion were supported by experiments on falling objects and the movements of the moon and planets.
- **Cell Theory / Germ Theory of Disease.** The theories that living things are composed of cells and that microscopic organisms cause disease are supported by observations through microscopes and by the experiments of scientists like Louis Pasteur.
- **Atomic Theory.** This theory is supported by how substances combine, evidence of radioactivity, the experiments of scientists like Rutherford who bombarded the atomic nucleus with particles, and the creation of nuclear weapons and reactors.
- **Theory of Evolution.** The theory that all life forms gradually evolved through natural selection is supported by scientific examination of fossils, DNA evidence, great diversity of life, the existence of extinct and present species, and the relationships among species.
- **Plate Tectonics Theory.** The theory that continents were once joined into a giant land mass and are moving apart is supported by the shapes of continents, the fossil remains of similar animals on distant continents, and space-based geodetic measurements.

SCORING GUIDELINES

Score point	Description
4 Points	The student identifies two important scientific theories and provides one example of evidence supporting each theory.
3 Points	The student identifies two important scientific theories and provides an example of evidence supporting one theory.
2 Points	The student identifies two major scientific theories but fails to or incorrectly provides evidence supporting each theory. **— OR —** The student identifies one major scientific theory and provides an example of evidence supporting this theory.
1 Point	The student identifies one major scientific theory.
0 Points	The student response does not meet the criteria to earn one point. The response indicates inadequate or no understanding of the task. It may only repeat information from the prompt or provide incorrect or irrelevant information. The student may have written on another topic or written, "I don't know."

Following are four sample responses to the extended-response question. Read each response carefully and then give it a score of **0 to 4 points**.

RESPONSE A:

Scientists often develop theories to explain what they observe. Two important scientific theories are the atomic theory and the germ theory of disease. Atomic theory states that all matter is composed of tiny particles called atoms. Atomic theory is supported by experiments in which scientists aimed radiation at atoms to discover their parts. The germ theory of disease states that many diseases are caused by microscopic organisms in the air. This theory was supported by the fact that the same organism can always be seen under a microscope when a person is affected by a particular disease.

Score

RESPONSE B:

Two important scientific theories are plate tectonics and the "Big Bang" theory.

Score

RESPONSE C:

- Newton's Theory of Gravity — All things fall.
- Cell Theory — All living things are made up of cells.

Score

RESPONSE D:

Two important scientific theories have been atomic theory and evolution. The theory of evolution states that living things gradually evolve based on natural selection. Organisms that are better suited to their environment are more likely to survive and have offspring. These offspring carry the favorable traits of their parents. Scientists have found many different types of evidence supporting evolutionary theory. For example, they find that fossils show how different species are linked to a common ancestor. DNA evidence also suggests many living species had a common ancestor.

Score

RESPONSE E:

The "Big Bang" theory is supported by evidence from many different scientists that proves that cells create a big bang when they hit into one another.

Score

Based on the rubric on page 28, the responses would most likely be scored as follows:

- ★ **Response A** correctly identifies two scientific theories — atomic theory and the germ theory of disease. In addition, the response provides evidence in support of both the atomic theory and cell theory. Therefore, since all *four* of the required points are found in this response, it should receive a score of **4**.
- ★ **Response B** correctly identifies *two* advances but *fails* to provide evidence in support of either theory. Even though the response is brief, it covers two of the required points. Therefore, it should receive a score of **2**.
- ★ **Response C** is written in bullet / note form but the student is not penalized for doing this. The response identifies Newton's theory of gravity and cell theory. However, it *mistakes* Newton's theory of gravity. In addition, the response provides *no evidence* in support of either theory. Therefore, this response should receive a score of 1.
- ★ **Response D** correctly identifies two theories and provides support for one of the theories (evolution). Therefore, since this response has three of the required points, it should receive a score of **3**.
- ★ **Response E** names one theory but gives wrong information as supporting evidence. This response demonstrates a *total lack of understanding* of the theory. It also *fails* to identify a second theory or to give any supporting evidence. Therefore, this response should receive a **0**.

As you can see, the most important part of answering any short-answer or extended-response question is reading the question **and** answering *all parts* of the question with **correct information**. The length of your answer will not determine your score. The amount of lines you write is far less important than providing correct information that answers the question.

THE NATURE OF SCIENCE

Science is a systematic method of investigating and explaining the natural world. The **OGT in Science** will test your knowledge of several fields of science. The first unit of this book reviewed common principles that relate to all these fields. To try to understand our natural world, scientists make observations, ask questions, develop theories, form and test hypotheses, and share ideas.

Lawrence Livermore Laboratories

In this unit, you will learn how scientific knowledge develops and changes, how scientists conduct investigations, and how scientific knowledge is closely linked to technology.

★ **Chapter 5: Scientific Knowledge and Inquiry**

This chapter focuses on the nature of scientific knowledge. You will learn how scientific inquiry guides the development of scientific knowledge, how science is based on models and theories that explain the natural world, and how scientists sometimes revise their theories based on experimentation and observation. You will also learn the importance of ethics in science and how science affects our everyday life.

★ **Chapter 6: Scientific Investigations**

This chapter looks more closely at the process of scientific investigation. You will learn how scientists form hypotheses and design experiments to test their ideas. You will also learn about the importance of laboratory safety.

★ **Chapter 7: Science and Technology**

In this chapter, you will learn how many recent advances in modern technology have been based on the discoveries of science. You will also learn how improvements in technology sometimes affect the development of science.

CHAPTER 5

SCIENTIFIC KNOWLEDGE AND INQUIRY

In this chapter, you will learn how scientists think. **Scientific inquiry** asks questions about the natural world. Answering these questions provides the basis for all **scientific knowledge.**

MAJOR IDEAS

A. **Scientific Inquiry** asks questions about the natural world in a systematic way and guides the growth of scientific knowledge.

B. **A scientific theory** attempts to explain data received from observing nature and conducting experiments. Models and mathematics help scientists develop theories, which guide scientific research and are continually revised.

C. **Scientific knowledge** explains the natural world in a logical way. The growth of scientific knowledge depends on communication among scientists.

D. **Ethical Behavior.** Scientists must behave ethically: they should report their results accurately and treat animal and human subjects with respect.

E. In modern society, every citizen should have some knowledge of science.

HOW SCIENTISTS EXPLAIN THE WORLD

Precise observations of the natural world and scientific experiments help scientists discover new information that enriches our scientific knowledge. Often an experiment raises new questions, leading to further inquiry. Eventually, experimentation leads scientists to revise their theories (possible explanations of what they observed).

★ **Scientific questions** must be specific, clear, and testable.

★ **Observations** are what we see, hear, smell, taste, and touch. We use our senses to observe.

★ **Data** are specific facts collected during an experiment or from nature. They are often measurements of *mass*, *temperature*, *time*, *speed* or *volume*.

★ **Scientific ideas** come from observations and data. They help us understand and explain what we observe. These ideas can be expressed as theories, models, or laws.

★ **Scientific knowledge** is the accumulation of all the scientific work that has been accomplished up to the present time. Scientific knowledge is developed through careful observation and experimentation, and builds on the work of others.

WHAT IS SCIENTIFIC INQUIRY?

Scientific inquiry often starts with a scientist's observations of the natural world. The scientist then asks questions about those observations. These questions lead to experiments. **Experiments** produce the data that make up the foundation of science. Scientists develop **theories** to explain large amounts of information and observations. Often a theory is based on a model, or simplified picture, of how things work. Scientists continually test, retest, and revise their theories.

Lawrence Livermore Laboratories

How do experiments build scientific knowledge?

The Role of Mathematics. Modern scientific theories are influenced by mathematics, a form of logical reasoning that can be applied to anything that can be measured. Once scientists develop a model for how something works, they apply mathematics to make predictions and test hypotheses. With a mathematical equation, scientists can precisely represent a complex relationship using only a few symbols.

Rigorous and Repeatable. Often, scientists test their theories through experiments that follow strict rules. They design an experiment in which they change only one variable and measure the effects. Scientific experiments must also be **repeatable**. This means that other scientists should be able to repeat the experiment and obtain similar results.

APPLYING WHAT YOU HAVE LEARNED

✦ Why is it important in the field of science for experiments to be repeatable?

SCIENTIFIC KNOWLEDGE GUIDES SCIENTIFIC INQUIRY

The Greek philosopher Aristotle wrote what is considered to be the first science textbook over 2,000 years ago. Since then, scientists have always looked for better ways to communicate. Scientific inquiry does not start from scratch: scientists build on the work of others.

WHERE SCIENTISTS FIND SCIENTIFIC INFORMATION

Good communication is one of the keys to good science. Scientists communicate with one another in a number of important ways:

- **Journals.** Scientists often publish articles about experiments in research journals. Before an article is published, a group of scientists read it, check how the experiment was done, and decide if its results are believable. The article is published and presented to others only if the scientists approve its contents.
- **Internet.** Scientists also publish articles on the Internet. Research journals often have websites. Computer search engines enable scientists to locate articles about experiments that interest them.
- **Meetings.** Scientists meet regularly to share the results of their experiments and discuss any new questions they may have. The information shared at these meetings is published and made available to all, including the public.

HOW SCIENTISTS DISCOVER NEW INFORMATION

Scientific work today has much in common with the way scientists have worked for the past 300 years. A scientist first finds out what others have done and what their findings mean. With this information, a scientist may choose to repeat an experiment to make sure the results come out the same, or the scientist may come up with a new question. This question often leads to yet another question. This leads to still further research, observation, and experimentation. Bit by bit, scientists increase our knowledge about the world.

Lawrence Livermore Laboratories

Before an experiment begins, a scientist finds out what others have done in that field.

SCIENTIFIC KNOWLEDGE CHANGES OVER TIME

Ideas about our world change as we learn more about it. For example, **Aristotle** once taught that Earth was the center of the universe. This model was generally accepted for almost 2,000 years. People who proposed that Earth revolved around the sun found their ideas rejected. Now we know they were correct. Scientific instruments enhance our powers of observation. They are regularly improved and new instruments are frequently invented. These instruments enable scientists to discover previously unknown facts about the world that increase our scientific knowledge. The more that scientists learn, the more they refine their explanations of what we see in the world and the better we understand how things really work.

OBSERVATIONS AND QUESTIONS GUIDE SCIENTIFIC INQUIRY

Scientists make observations about the world and continually ask questions based on their observations. Their desire to understand the natural world leads scientists to conduct scientific inquiry. This can be seen in the cases of Sir Isaac Newton and Louis Pasteur.

★ **Sir Isaac Newton.** Watching apples fall from trees, Newton wondered if the same force held the moon in its orbit? He studied the movements of the planets in the solar system, and read the experiments of other scientists like Galileo. Newton then asked questions and designed experiments of his own to find out more. Over time, he noticed patterns. He observed that the orbit of the moon could be related to the path of a cannon ball shot from the Earth's surface. Newton used mathematics to develop laws that explain these movements.

Library of Congress

Sir Isaac Newton

★ **Louis Pasteur.** Louis Pasteur wanted to know why milk got sour and made people sick. He studied rancid foods under a microscope and discovered that bacteria were always present. Pasteur theorized that bacteria made food spoil. He experimented and found the bacteria died when foods were heated to a high temperature. This heating process is called *pasteurization*. The milk we buy today has been pasteurized so that we can drink it safely.

Questions lead to experiments and further observations. At the end of an experiment, a scientist must look at all the observations made and the data gathered. The scientist must interpret, or explain, the data. Interpreting data often leads to a theory or general explanation of what happens in nature. Newton asked why the moon follows a particular path around Earth. This question led to the **Theory of Universal Gravitation** (*gravity*). Pasteur asked how milk spoiled. This question led to a better understanding of microorganisms and safer milk.

APPLYING WHAT YOU HAVE LEARNED

✦ What other examples can you think of where new observations led to new scientific ideas?

IDEAS GUIDE SCIENTIFIC INQUIRY

A "big idea" in science is called a **theory** — it attempts to explain all of the observations and data collected through experimentation and observation of nature.

A theory acts as a framework to hold related information together. For example, Newton studied the movement of planets and how things fall on Earth. Then he came up with the theory of universal gravity. His theory helps explain what we see, and allows us to predict what will happen. We use that information to accomplish many tasks, including launching rockets into space.

Theories also guide research. Scientists can test a theory by making predictions based on the theory and then seeing if their predictions come true.

Another way to test a theory is to perform new experiments to see if the theory can explain the results. Theories must withstand repeated testing or they must be changed or discarded. If an experiment results in new information, scientists sometimes change the theory to fit the new evidence. New data will either support the theory, require that the theory be revised, or cause the theory to be rejected. For example, when **Albert Einstein** studied how light travels, he discovered new things that led him to revise Newton's laws of motion.

Albert Einstein revised many existing theories

APPLYING WHAT YOU HAVE LEARNED

✦ Why is the constant revision of theories necessary for scientific progress?

SOME KEY SCIENTIFIC THEORIES*

Earth and Space Sciences	★ Copernican Heliocentric Theory ★ Plate Tectonic Theory
Physical Sciences	★ Theory of Universal Gravity ★ Newton's Laws of Motion ★ Atomic Theory
Life Sciences	★ Cell Theory ★ Germ Theory ★ Theory of Evolution

**Each of these scientific theories will be discussed in a later chapter.*

THE ETHICS OF SCIENCE

Science depends on scientists' accepting and practicing an ethical code. This means that scientists have to perform their research honestly and report their results accurately. A scientist cannot influence experimental results in any way. This **code of ethics** is required to reduce bias, or prejudice.

"Peer review" often acts as an ethical safeguard in science. Articles are reviewed by a scientist's peers, or equals, before being published. Each article must accurately report procedures and results, so that others can repeat and verify the work.

LIMITS ON WHAT SCIENTISTS CAN DO

Scientists cannot do something just because it is possible. They must operate in a society that accepts their work as contributing to a better quality of life, not harming life. Society funds scientific research. These funds come from the government, from businesses investing in products like medicines, and from private grants. Society often has strong ideas about what science should and should not do. Some of these expectations are formalized as laws, while others are established as policies.

Science Photo Library

The world's first cloned sheep and the scientist who created her. Cloning raised many ethical questions about animal research.

HUMANS AND ANIMALS IN SCIENTIFIC RESEARCH

All scientists involved in research must follow strict laws and guidelines. Committees of government scientists and others can stop research at any time if they think it is unethical. Using human subjects in research is desirable, especially when new drugs are being developed. A new drug cannot be offered to millions of people unless scientists are certain it is safe and effective. But such research must meet strict standards. To participate in scientific research as a subject, a human must volunteer and give **informed consent**, usually in writing. The researcher must fully explain the risks and the subject must understand them.

Research with animal subjects is also strictly regulated and must be conducted according to professional standards and laws. For example, animals must be kept in sanitary conditions. Oversight committees make sure that researchers follow these standards and laws.

APPLYING WHAT YOU HAVE LEARNED

- ✦ Why are ethical guidelines needed by scientists?
- ✦ How active should the government be in imposing ethical requirements?
- ✦ What ethical rules would you impose on scientific research?

THE IMPORTANCE OF SCIENCE TO EVERYDAY LIFE

A person who has **scientific literacy** knows about the contributions science has made to society. Computers, televisions, automobiles, airplanes, and electricity all remind us of the importance of science in our daily lives. We can eat foods grown in distant places because refrigerated trucks keep them from decaying while they are transported. We live longer and healthier lives because we have better medicines and sanitation than people in the past. People of all races, genders, and backgrounds — both in and out of science — have contributed to these advances. You should realize that scientific literacy is important to being a good citizen.

Science makes the foods we eat healthier.

Examples of How Science Has Changed Our Lives

- ★ **Biology.** We now know that germs cause many diseases, and scientists are using biotechnology to find new cures.
- ★ **Chemistry.** We enjoy the use of many new materials, such as plastics and synthetic fibers, based on scientific knowledge.
- ★ **Physics.** Physicists helped develop x-rays, television, nuclear energy, and nanotechnology.

APPLYING WHAT YOU HAVE LEARNED

- ✦ What do you think has been the greatest contribution of science to most average citizens? Explain your choice.

CITIZENS VOTE AND PAY TO KEEP SCIENCE WORKING

Ordinary citizens make important decisions about science. Voters decide what scientific work will be funded, how scientists must operate, and how scientific advances will be used. If voters do not know enough about science, they will not make the best decisions.

SCIENCE IN THE CLASSROOM

Today's science students are tomorrow's voters and scientists. Many students learn most of their science in high school. The reasoning and thinking skills learned in science are used in other subjects. The knowledge, skills and interests learned by students in science classes often plant the seeds for a future career in science. Students may become scientists, doctors, nurses, or laboratory technicians.

WHAT YOU SHOULD KNOW

Make sure you know how:

- ★ scientific inquiry guides scientists' exploration of the natural world.
- ★ scientific theories help explain and predict the natural world.
- ★ scientific knowledge develops over time based on scientific evidence.
- ★ ethical guidelines influence the work of scientists.
- ★ scientific ideas and skills apply to present-day careers and society.

CHAPTER STUDY CARDS

Scientific Inquiry

★ **Scientific Inquiry.** Scientists ask questions about what they observe in the natural world. Inquiry guides research.

★ **Theory.** A possible explanation of observations and data, which is tested and revised. Theories guide research and build scientific knowledge.

★ **Experiments.** These produce much of the data that make up the foundation of science.

Scientific Knowledge

★ Changes over time as scientists make further observations, ask new questions, conduct new experiments, and revise theories.

★ **Case Studies.**

- **Aristotle** thought that the sun moved around the Earth. Later, this idea was proven wrong by other scientists.
- **Isaac Newton** studied planets and how objects fall to Earth. From this, Newton developed laws of motion and gravity.
- **Albert Einstein** revised Newton's laws.

APPLYING WHAT YOU HAVE LEARNED

✦ In addition to the two study cards above, create two additional cards of your own using the following headings:

- **Ethical Standards in Science**
- **Science and Citizenship**

CHECKING YOUR UNDERSTANDING

1 A scientist ignores some of her experimental results since they do not support a hypothesis she is testing. She submits the results of her experiment to a scientific journal for publication. What ethical argument could the editor give for refusing to publish her report if he learns what she has done?

SW: C* 10–4

A. Readers would not be interested in reading about her experiment.

B. The editor preferred articles by companies that paid to advertise in the journal.

C. Other scientists were testing the same hypothesis.

D. The author had provided biased results from her experiment.

♦ Examine the Question
♦ Recall What You Know
♦ Apply What You Know

* Every question in this book is identified by its **standard**, **benchmark**, and **grade level indicator**. This question tests **Scientific Ways of Knowing**, **Benchmark** C, and the **4TH grade level indicator** for the tenth grade. A complete list of benchmarks and grade level indicators can be found on page 212.

This question examines your understanding of the ethical practices of scientists. Recall the ethical duties required of scientists reporting an experiment. Then apply this knowledge to the question. The correct answer is D.

Now try answering some additional questions on your own about scientific inquiry and knowledge.

2 **Scientific knowledge is based on the development of theories that attempt to explain the natural world. Give one example of a theory and show how it was revised based on later experiments or observations.** (*2 points*)

SW: A 10–3

3 **A scientist is studying the effects of a proposed new drug on human subjects. Which of the following practices by the scientist would be considered unethical?**

- ♦ Examine the Question
- ♦ Recall What You Know
- ♦ Apply What You Know

SW: C 9–4

A. The scientist makes a statistical study of people who might be helped by the drug.

B. The scientist researches what other scientists have discovered about similar drugs.

C. The scientist studies the drugs using a high-powered electron microscope.

D. The scientist injects human subjects with the drug without fully explaining the risks.

Use the following statements to answer the question:

- **Galileo's experiments found that a body in motion tended to stay in motion.**
- **Newton's first law of motion stated that a body at rest would remain at rest and a body in motion would stay in motion at the same speed and direction unless another force acted on it.**

4 **What conclusion can be drawn from these statements?**

SW: A 9–7

A. Scientists frequently build on the work of earlier scientists.

B. Scientific investigation always leads to improvements in society.

C. Experimentation is more important to science than observation.

D. The work of earlier scientists seldom influences later scientists.

5 **How do experiments differ from theories?**

SW: B 9–6

A. Experiments are based on observations; theories are based on data.

B. Experiments look at the natural world; theories generally do not.

C. Experiments produce data; theories provide explanations of data.

D. Experiments can be verified; theories cannot be proven or disproven.

CHAPTER 6

CONDUCTING SCIENTIFIC INVESTIGATIONS

In this chapter, you will learn about the process of scientific investigation and its main steps, including the roles of evidence and prediction.

MAJOR IDEAS

Use the following steps when conducting an experiment to test a hypothesis:

★ Identify the independent and dependent variable in your experimental design. Hold other variables constant during the experiment. Make sure you can measure your results. Be sure your design tests your hypothesis.

★ Take safety into account before beginning any experiment. Check the Material Safety Data Sheet (MSDS) for any chemical you may use. Have safety equipment on hand like goggles, gloves, and a fire extinguisher.

★ Use lab instruments to take accurate and precise measurements. The results you record should be no more precise than your least precise measurement.

★ Relate the results to your hypothesis in your conclusion. Your results should lead you to support, reject or change your hypothesis.

★ Communicate your results to others to learn their interpretations and to contribute to scientific knowledge.

THE STEPS OF A SCIENTIFIC INVESTIGATION

Scientists use special methods to investigate the natural world. They try to control conditions to test their ideas in an experiment. Specific procedures vary from experiment to experiment, but good scientific work always follows a logical, step-by-step approach.

While there is no single method of scientific investigation, many scientists use the following steps:

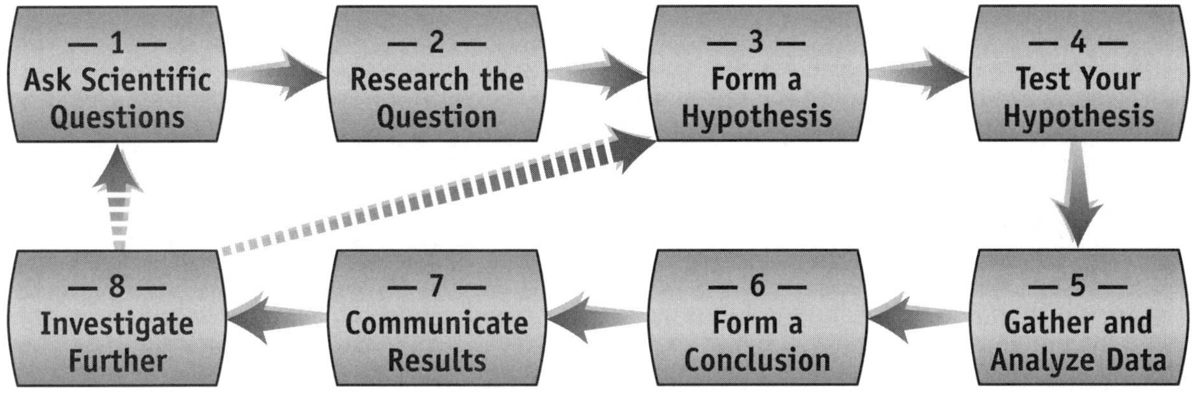

As the dotted arrows show, results often suggest new questions or approaches to be followed.

ASK SCIENTIFIC QUESTIONS

Scientists develop questions based on their observations of nature. Then they refine these questions for scientific investigation. Vague questions cannot be answered by scientific investigation. Scientific questions must be ***clear***, ***specific***, and ***testable***.

★ **Clear.** A clear research question can be easily understood. For example, *What is the effect of fertilizers on plants*? is ***not*** clear. The reader will wonder which fertilizers, what kind of effect, and what kind of plants the question refers to.

★ **Specific.** A specific research question identifies exactly what will be tested in an experiment. For example: *How does nitrogen-based fertilizer affect the growth of bean plants*? This question is more specific than the one above: it states what kind of fertilizer is used, what effect is observed, and what kind of plant is studied.

★ **Testable.** An ideal research question can be tested in an experiment. A researcher isolates one variable in the experiment (*fertilizer*) to see how changes in that variable affect another variable (*plant growth*).

RESEARCH THE QUESTION

Good scientists consult a variety of references and sources to find what is already known about a subject. A scientist may start with general references (*textbooks and encyclopedias*), and then use more specific ones, such as journal articles, websites, and other scientists.

★ **Look for the Question.** A scientist needs to know if anyone else has researched the same question. If so, the scientist must gather as much of this research as possible. The scientist may then change his or her question based on this research.

★ **Look at Similar Questions.** A scientist also needs to know what research has been done on similar questions, which may also apply to the question the scientist is investigating.

★ **Look beyond the Question.** A scientist tries to think about the results the experiment will produce and what questions to ask next. Looking beyond the original question often helps the researcher to more fully understand that question.

FORM A HYPOTHESIS

A **hypothesis** is an educated guess that attempts to answer the question under investigation. For example, a scientist may make the hypothesis that *nitrogen-based fertilizers help bean plants to grow*. This hypothesis may turn out to be right or wrong. In science, proving that a hypothesis is wrong can be just as valuable as confirming the hypothesis.

APPLYING WHAT YOU HAVE LEARNED

✦ Why do you think it is just as important to prove that a hypothesis is wrong as to prove it is valid?

✦ Think of an experiment you did in science class this year. What hypothesis did that experiment test?

TEST YOUR HYPOTHESIS

An experiment creates controlled conditions to test the hypothesis and answer the proposed question. Often the experimenter turns the hypothesis into a prediction: *if bean plants are given nitrogen-based fertilizer, then they will grow faster than bean plants without the same fertilizer*. The **experimental design** is the roadmap of the experiment. An experimenter must have a clear understanding of the **variables** (*things that can change*) in an experiment. Some variables can be manipulated (*controlled*) by the experimenter. For example, the experimenter in a plant experiment can decide how much fertilizer to use, how much water to use, and how long to keep the plants in sunlight. Other variables cannot be manipulated. The experimenter cannot directly control what happens. The height that each plant grows is an example of this second kind of variable.

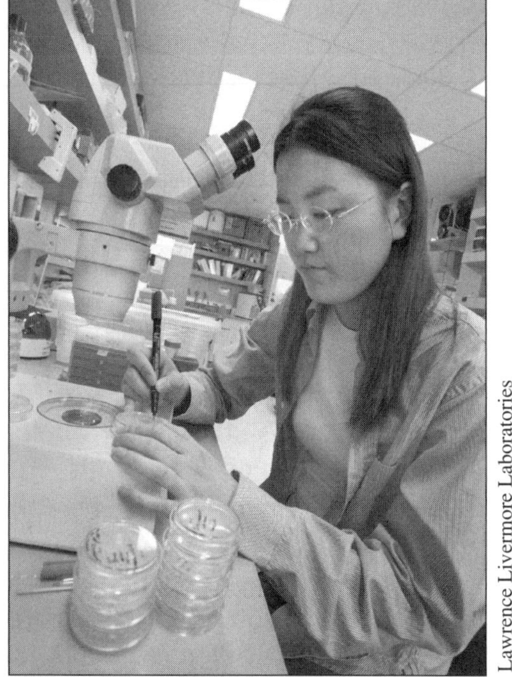

Lawrence Livermore Laboratories

AN EXPERIMENT HAS DIFFERENT KINDS OF VARIABLES

★ **Independent (*manipulated*) Variable.** A variable an experimenter changes to find out how a change in this variable affects other variables in the experiment (*e.g., how much fertilizer each plant gets*).

★ **Dependent (*responding*) Variable.** A variable that changes as the result of a change to an independent variable in the experiment (*e.g., how much each plant grows*).

★ **Variables Held Constant.** All the variables in an experiment that must be kept the same (*e.g., the type of plant, the amount of sunlight*).

Controlling Variables. In a controlled experiment, the experimenter changes only *one variable* at a time. For example, in an experiment to test nitrogen-based fertilizer, a scientist will use only one kind of bean plant, with all plants as close to the same size as possible. The experimenter will provide the same amount of water at the same time to all the plants. The plants will all be exposed to the same amount of sunlight. Everything will be as close to the same as possible except for the tested variable.

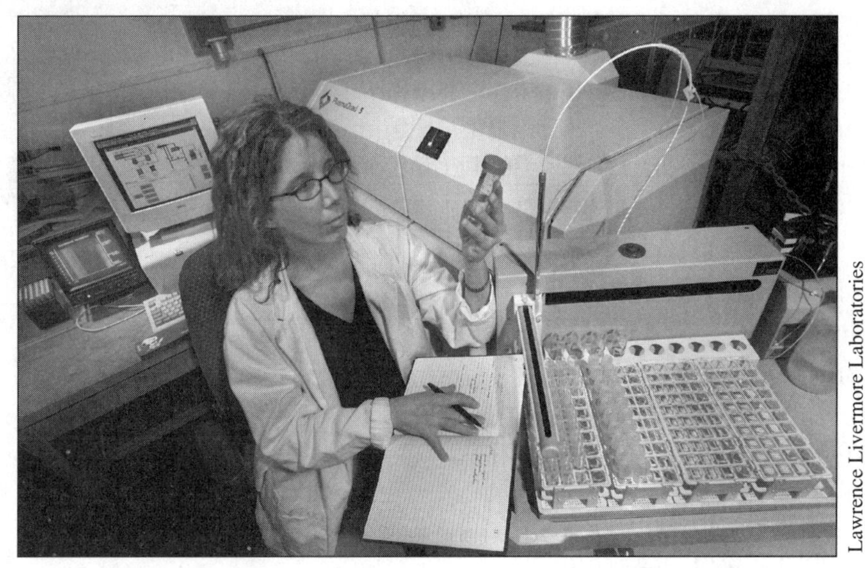
Lawrence Livermore Laboratories

Comparing results between groups is essential in an experiment.

Experimental Group vs. Control Group. The only thing that changes in this experiment is whether a plant gets fertilizer. Some plants will be given fertilizer; a second group of plants, known as the **control group**, will receive none. The scientist then can compare the results of the experimental group with the control group. The experimenter can be fairly confident that any significant differences in plant growth probably result from the plants' getting or not getting fertilizer. By running several **trials**, the experimenter can increase confidence in the results.

Designing Procedures. After the variables are determined, the researcher must fill in the details of the experimental design. The designer should identify all necessary materials and list all the steps that need to be performed. The researcher should try to think of all the things that can go wrong. The researcher must make sure that all variables except the independent variable are controlled, and should discuss the design with other scientists.

Changing the Design. Experimental designs are continually evaluated and sometimes changed. If something in a design does not work, the researcher must change it or the results of an experiment will not be valid.

ELEMENTS OF A GOOD EXPERIMENTAL DESIGN

- ★ Variables can be measured with instruments.
- ★ There should be several trials.
- ★ All variables need to be identified, and only one is experimental.
- ★ All necessary materials are listed, including amounts or sizes needed.

The goal of an experiment is to obtain valid and reliable results. If a researcher sees that such results are not possible, the design must be changed. A research design may also change to answer new questions. For example, a researcher may want to study the effects of *different kinds* of fertilizers on bean plants. Ten plants may get one kind of fertilizer, ten may get another type, and ten may get a third type. A final group of ten plants would act as a control group and receive no fertilizer.

Science Photo Library

Students work together on a class experiment.

Safety Considerations. Safety is important in any experiment. It must be considered before the experiment begins, and precautions must be included in the experimental design. For example, will hazardous chemicals be used? Do the chemicals produce dangerous fumes? How will chemicals be disposed of? Are there any biological hazards and how should they be handled? Are any poisonous plants used? Will open flames be used? Is a fire extinguisher nearby? How should eyes be protected? Is there a risk of fire or explosions?

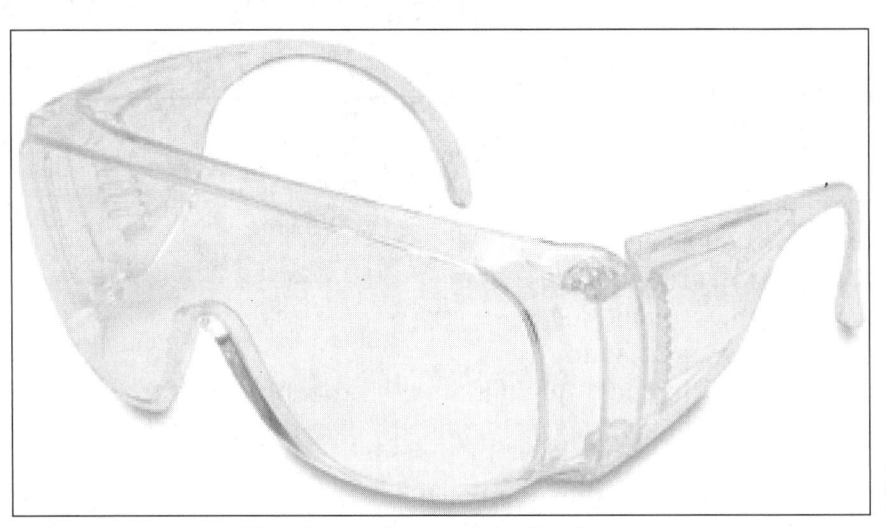

Goggles are an essential safety feature

Material Safety Data Sheets. Before using any chemicals, scientists consult the Material Safety Data Sheet (MSDS) for that chemical. The MSDS tells what steps to take when using that chemical. It also tells what to do in case of accidental eye contact, if the chemical is swallowed or breathed in, if a fire breaks out, or if some other emergency occurs.

SOME COMMON LABORATORY SAFETY RULES

- Wear safety goggles, a laboratory apron, and gloves for most experiments.
- Read all chemical labels and safety symbols. Be sure to check the MSDS.
- Know where to wash your hands or eyes and where the fire extinguisher is located.
- Wash hands with soap before and after all experiments.

GATHER AND ANALYZE DATA

Data are pieces of information gathered during an experiment. Data can be in the form of measurements or observations. A scientist uses a notebook as a diary or log during an experiment. When something is done, observed, or measured, an entry is made in the notebook. All the information entered is dated. A good scientist will not erase in a lab notebook. Errors are crossed out. Everything ever written in a lab notebook needs to be visible.

Taking Measurements. The measurements recorded in the laboratory notebook come from using appropriate tools. Measurements are made using the metric (SI) system. For example, you could use a meter stick to measure the height of a bean plant in centimeters. All measurements should be taken several times and recorded with their units in the log. When using a mercury or alcohol thermometer, keep your eye at the same level as the colored liquid to read the degrees. When using a graduated cylinder, you should be aware that liquids create a concave, or curved, surface. Measure the amount of liquid from the **lowest** point of the curve.

Several graduated cylinders

COMMON MEASUREMENTS USED IN EXPERIMENTS

Length		Mass		Volume	
centimeter	cm	milligram	mg	milliliter	mL
meter	m	gram	g	liter	L
kilometer	km	kilogram	kg	cubic centimeter	cm^3

Precision refers to how exact a measurement is. A tool measuring centimeters is more precise than one measuring meters. **Accuracy** refers to whether a measurement is correct or not. A scientist might use precise measuring tools but make an error in taking measurements.

Significant Figures. The data a scientist records cannot be more precise than the measuring tools used. A scientist therefore often uses round numbers. For example, a scientist may need to determine a sample's density. The mass is measured as 16.8 grams and the volume as 3.9 cubic centimeters. Dividing the numbers with a calculator gives the density as 4.3076923 g/cm^3. A scientist will round this answer to the same number of figures as the *least precise measurement* (3.9). The density is therefore rounded to 4.3 grams/cubic centimeters.

Observation vs. Inference. An experimenter should know the difference between an observation and an inference.

★ **Observations.** An **observation** is made by one of the senses and includes what an experimenter *sees*, *hears*, *smells*, *feels*, or *tastes*.

★ **Inference.** An **inference** is something an experimenter *concludes* from an observation. An observation is that the water in a pot is at 90° Celsius. An inference is that the water is close to boiling.

Looking for Patterns. Once a scientist has gathered data from an experiment, this data must be analyzed. A scientist looks for patterns in the data. The best way to do this is often to draw a picture from the data. **Graphs** are pictures of data. With a good graph, a researcher can often see a trend (*or lack of a trend*) between two variables with one glance. A **bar graph** is often useful for comparing items. A **line graph** is best for showing different measurements of the same item over time.

DRAW CONCLUSIONS

Scientists draw conclusions based on the results of an experiment. These conclusions should relate back to the hypothesis.

WHAT MAKES A GOOD CONCLUSION

★ **Logical.** A conclusion must be logical and correctly report and interpret the data.

★ **Supported.** It must be supported by scientific knowledge and evidence from the investigation.

★ **Relevant.** The conclusion should support or reject the hypothesis, or suggest changes in the hypothesis for further study.

Scientific theories are often based on conclusions that come from scientific investigations. Every good scientific investigation supports, rejects, or changes a theory.

COMMUNICATE RESULTS

Once a scientist completes an experiment, the results must be communicated to others, usually in a written report or article. Sometimes scientists give oral presentations at meetings. A good presentation uses clear language, includes accurate data and graphs, and includes the methods and procedures followed so that other scientists can repeat the experiment and verify the results. This encourages others to investigate and build on what was learned.

Sometimes disagreements may develop over the experimenter's conclusions or interpretations of the data. This process of questioning, disagreement and debate increases scientific knowledge. Good criticism can help refine an investigator's questions or procedures. Scientists may not always reach the same conclusions, but good scientists all accept the value of scientific investigation and open debate.

INVESTIGATE FURTHER

After drawing conclusions and communicating results, a scientist often thinks of new questions for further research.

CASE STUDY: DESIGNING AND CONDUCTING AN EXPERIMENT

Read the following description of how a student conducts a scientific investigation. This student works in much the same way as a professional scientist would.

ASK A QUESTION

A student is interested in fashion. She lives in a hot climate and is aware that some fabrics make her feel cooler than others. The student decides to investigate the fabrics used to make clothes and to design an experiment that measures the temperature under each fabric when it is exposed to heat.

RESEARCH THE TOPIC

She looks in encyclopedias, scientific journals and the Internet for information about various fabrics and their appropriateness for use in different temperatures.

FORM A HYPOTHESIS

A hypothesis for this experiment might be: *Clothes made from white cotton transmit less heat in hot weather than clothes made from other fabrics.*

DESIGN THE EXPERIMENT

The student designs an experimental model of a person wearing clothes made from various fabrics. A thermometer represents the person, a heat lamp represents the sun, and fabric samples represent the clothes. The student selects 10 different white fabrics to test. Each sample is the same size. Each sample is examined under a microscope to see how far apart its threads are. The student creates a notebook to record all data after the experiment begins.

EXPERIMENTAL VARIABLES

- ★ **Independent variable.** The type of fabric (*including its thread count*) will be the independent variable in the experiment.
- ★ **Dependent Variable.** The temperature of the thermometer will be the dependent variable in the experiment.
- ★ **Variables Held Constant.** The size of the sample, the distance of the lamp and amount of heat, and the time heat is applied are variables that do not change.

CONDUCT THE EXPERIMENT

The student gathers the materials and begins the trials. She follows strict safety precautions with the heat lamp. The same heat lamp is used for the same amount of time over each sample. The thermometer is placed behind each sample in the same position and at the same distance. Each fabric is exposed to the heat lamp for 15 minutes per trial. She records the temperatures during five trials.

EVALUATE THE RESULTS

She records the results of her observations in a log. The student compares the average temperatures of the thermometer behind each fabric. After all the results have been recorded, she concludes that the white cotton fabric transmitted the least heat. This data confirms her hypothesis.

COMMUNICATE RESULTS

To communicate the results of this experiment she creates a bar graph. She lists each type of fabric on the bottom axis with its average temperature under the heat lamp shown on the vertical axis.

WHAT YOU SHOULD KNOW

★ You should know how scientists design and conduct experiments, maintain safety, take precise measurements, and analyze results.

★ You should know the difference between independent, dependent, and constant variables in a controlled experiment.

★ You should know why it is important that scientists communicate their findings.

CHAPTER STUDY CARDS

Steps in a Scientific Investigation

★ **Ask Scientific Questions.** Develop specific questions based on observations.
★ **Research the Question.**
★ **Form a Hypothesis.** Make an educated guess to answer the question.
★ **Test the Hypothesis.** The experiment alters an **independent variable** to see the effects on the **dependent variable.**
★ **Gather and Analyze Data.**
★ **Draw Conclusions.**
★ **Communicate Results.**
★ **Investigate Further.**

Laboratory Experiments

★ **Laboratory Tools.** These should be able to provide precise measurements. These tools include a balance, meter stick, thermometer, graduated cylinder, pH meter.
★ **Laboratory Safety.** Safety should always be an important concern in all experiments.
★ **Experimental Group vs. Control Group.** The "control group" is not subject to changes in the "independent variable." Changes in the experimental group are then contrasted with the control group.

APPLYING WHAT YOU HAVE LEARNED

✦ Make two cards of your own on the following topics:

- **Types of Variables**
- **Laboratory Safety**

CHECKING YOUR UNDERSTANDING

The diagram below shows two germinating corn seeds that have been placed in identical bottles and kept in the dark. Bottle A will be rotated 90° each day for the next 6 days. Bottle B will not be rotated.

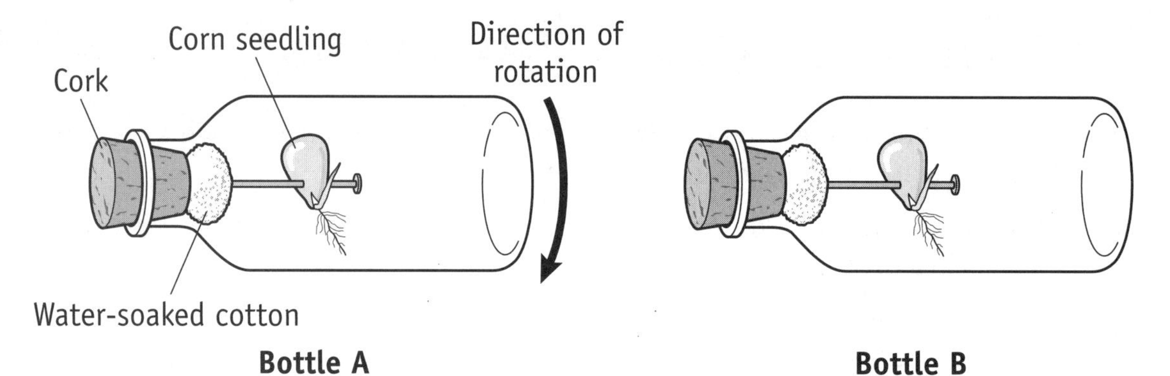

1 Which hypothesis is tested in this experiment?

A. Water is needed for proper plant growth.
B. Gravity affects plant growth.
C. Enzymes promote seed development.
D. The amount of light received affects chlorophyll production.

- Examine the Question
- Recall What You Know
- Apply What You Know

SI: A 10–5

This question examines your understanding of the methods of scientific investigation. You should recall that an experiment usually tests the effects of changing one "independent" variable. Here, only one variable is changed: whether the bottle is turned or not. There is no light in the room, and there is no difference in the water or enzymes. Therefore, the answer is B. The experiment tests the effects of gravity.

Now try answering some questions on your own about scientific investigation.

2 Which of the following represents a testable hypothesis?

SI: A 10–5

A. Environmental conditions affect germination.
B. Boil 100 milliliters of water, let it cool, then add 10 seeds to the water.
C. There is less sunlight in the bottom of a pond where the water is deeper.
D. A lamp, two beakers, and elodea plants are needed for investigation.

3 How does the control group in an experiment differ from other groups in the same experiment?

A. It tests a different hypothesis.
B. It has more variables.
C. It differs in the one variable being tested.
D. It utilizes a different method of data collection.

♦ Examine the Question
♦ Recall What You Know
♦ Apply What You Know

SI: A 9–6

4 A marine biologist carefully measures the dimensions of a rectangular aquarium to be used in an experiment on dissolved oxygen and microscopic forms of marine life. The dimensions of the aquarium are recorded in her laboratory notebook as follows:

Lengh = 50 cm
Width = 10.52 cm
Height = 8. cm

Next, she calculates the volume of the aquarium. Which of the following identifies the volume with the correct number of significant figures based on her recorded measurements?

A. 4,208 cm^3
B. 4,208.0 cm^3
C. 4,208.00 cm^3
D. 4,208.000 cm^3

SI: A 9–4

5 An experimental design included references to prior experiments, a list of materials and equipment, and detailed step-by-step procedures. What else should be included before the experiment can be started?

A. a set of data
B. an inference based on results
C. a conclusion based on the data
D. a list of safety precautions to use

SI: A 10–1

Use the information in the passage below to answer the following question:

You head the research division of the Leafy Lettuce Company. Your company is experimenting with growing lettuce using hydroponic technology, which involves growing plants in containers of growth solution in a greenhouse. No soil is used. The growth solution that the company uses contains water, nitrogen, and phosphorus. The company wants to know if adding iron to this formula will improve the growth.

6 State a hypothesis for an experiment on the effect of adding iron to the Leafy Lettuce Company's formula to improve growth. Then identify two factors that must be kept the same in both the experimental and control groups. (*4 points*)

SI: A 9–3

7 The piece of laboratory equipment on the right can be used to measure the

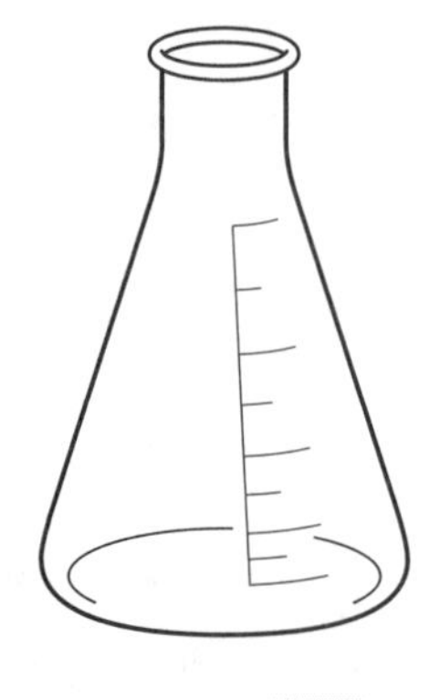

A. temperature of a flame
B. volume of a liquid
C. mass of a powder
D. pressure exerted by a gas

- ♦ Examine the Question
- ♦ Recall What You Know
- ♦ Apply What You Know

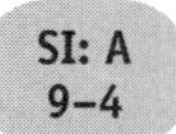

8 When heating a solution in a test tube, a student should be sure to

A. point the test tube in any direction
B. hold the test tube with two fingers
C. cork the test tube
D. wear goggles

SI: A 9–2

9 Which of the following pieces of lab equipment can safely be used to clamp a test tube before safely heating it?

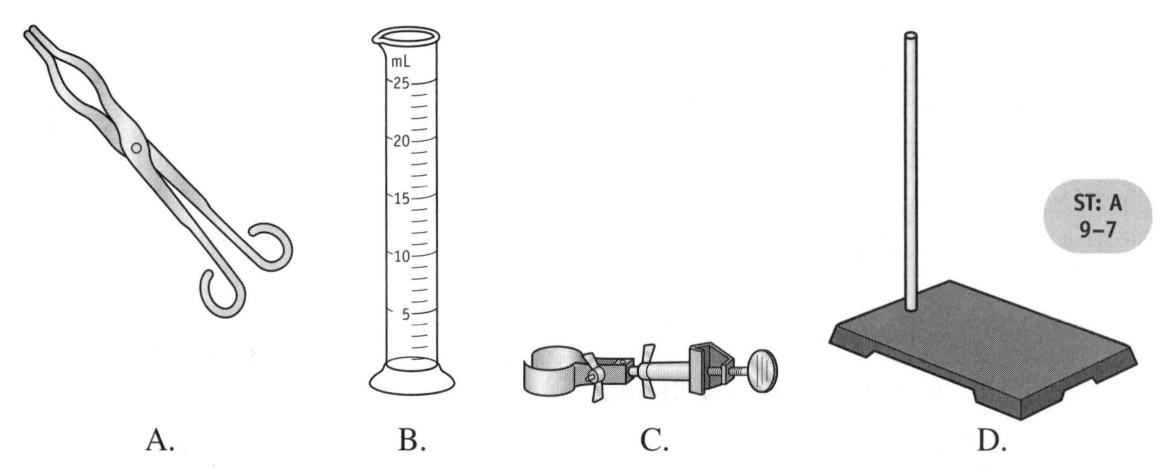

A. B. C. D.

ST: A 9–7

10 A graduated cylinder is filled with water to 28 mL. A rock is placed in the cylinder, and the water rises to the 57mL mark. What is the volume of the rock?

A. 57 mL
B. 85 mL
C. 29 mL
D. cannot be determined

SI: A 9–6

11 The following balance measures mass in grams. What mass is shown with correct precision?

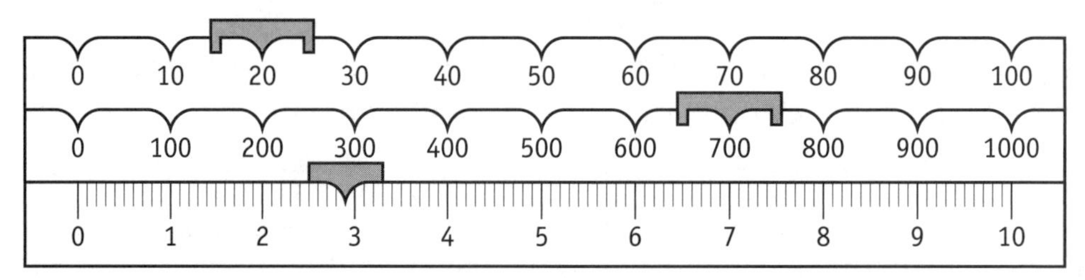

A. 722.90 g
B. 722.9 g
C. 720 g
D. 723 g

SI: A 9–4

CHAPTER 7

SCIENCE AND TECHNOLOGY

Technology is the application of scientific knowledge to develop tools, materials, and systems that help humans meet their needs. In this chapter, you will learn how scientists use technology to respond to the needs of society.

MAJOR IDEAS

A. Technology often applies scientific knowledge to human needs. For example, scientific research led to jet travel, synthetic fibers, microwaves, and antibiotics.

B. The processes of technological design and scientific investigation are often quite similar, except that technological design is generally aimed at meeting a particular need.

C. In deciding whether to adopt a new technological design, societies must consider costs of manufacture, how easy the new technology will be to distribute and operate, safety to users, and potential effects on the environment.

D. Two emerging fields that combine science and technology are biotechnology and nanotechnology.

HOW TECHNOLOGY MEETS OUR NEEDS

Scientific inquiry is driven by the desire to understand the natural world around us. In contrast, **technology** is driven by the desire to meet human needs and to solve human problems. Nevertheless, both science and technology are closely related. In fact, many of the modern technological wonders we use in everyday life resulted from scientific work:

★ Scientists discovered that the electrical conductivity of the element selenium changed with the amount of light. This discovery led to the invention of a practical television set.

★ A scientist working with an electron tube noticed that the candy bar in his pocket melted. He used the idea behind this discovery to build the first microwave oven.

- ★ A chemist was combining chemicals in a beaker. He pulled out a stirring rod a new fiber clinging to it — and discovered nylon.
- ★ The discovery that sunlight produced a weak electric current when it fell on certain materials led to solar cells.
- ★ A research team found out how to view a virus in three dimensions. Their work is leading to the development of new drugs to fight colds and other viral infections.

THE TECHNOLOGICAL DESIGN PROCESS

Most new technology results from a process, just as most scientific discoveries result from the systematic methods of scientific investigation. In many ways, the process of technological design is similar to the process of scientific inquiry. Technological designs are continually tested, adapted and refined — just like scientific ideas. The steps of the technological design process can vary just as they may in any scientific inquiry.

STEPS IN TECHNOLOGICAL DESIGN

- ★ **Identify the Problem.** Identify a need or problem.
- ★ **Research the Problem.** Research what is known about the problem, what has already been done to address the problem, and what limits must be considered (*e.g., environmental concerns, weight problems, costs*).
- ★ **Design a Product.** Design a new product or process to meet the need or to solve the problem, making use of current scientific knowledge and existing technologies.
- ★ **Build a Sample.** Build a sample (*or model*) of the design and test it to see if it works. Redesign the sample as necessary and keep testing.
- ★ **Test and Evaluate.** Build a final design sample of the technology, and test and evaluate it. Accept the final design, reject it as unworkable, or redesign and again retest.

Once a design is accepted, researchers still have more decisions to make. The design must be practical. Modern products must be mass-produced, affordable, safe to use, and last long enough to be worth the cost. The design must also be desirable. People must want to buy the new product, process or system.

HOW A NEED BECAME A NEW TECHNOLOGY

Zippers are not the most sophisticated examples of technology, but it would be hard to live without them. In the late 1800s, people had to lace their high boots with buttonhooks. The lacing took a long time, and people had to bend over for the whole procedure. Whitcomb Judson, a Chicago engineer, had an idea for a faster, easier way to close boots. He researched fasteners and found that nothing even remotely like his idea existed.

Judson then made a prototype of his zipper, calling it a "clasp-locker." He received a patent in 1893. But no one was interested in buying his complicated-looking product. The United States Postal Service bought some bags with a zipper on them, but found that the clasp-locker often jammed. It was left to another engineer, Gideon Sundback, to produce a smaller, lighter, and more reliable zipper than Judson's product. Sundback's product looked more like our modern zipper.

Gideon Sundback

By 1920, the zipper began appearing on clothes. But early zippers were unpopular. People were so confused by them that clothes with zippers came with an instruction book on how to use and care for them. Early zippers rusted easily. People had to unstitch a zipper from a piece of clothing before washing it and then sew it back in after the clothing had dried. In 1923, the B.F. Goodrich Company put zippers on rubber galoshes. These zippers were more reliable and didn't rust so easily. The company also coined the name "zipper." This combination of improved performance and a catchy name made zippers popular. By the late 1920s, people finally began to accept zippers.

Notice how Judson identified a need, researched for possible solutions, and made a product. His market test of the product, however, was not successful. Sundback made improvements to Judson's product and successfully tested his improved design. The B.F. Goodrich Company improved Sundback's product further and expanded the market to the general public. The more thought that is put into a technology and the more it is tested, the greater the chance that the technology will succeed.

APPLYING WHAT YOU HAVE LEARNED

✦ How does the process of technological development compare with the growth of scientific knowledge? Identify both similarities and differences.

WHAT RESEARCHERS THINK ABOUT WHEN EVALUATING A NEW TECHNOLOGY

Certain questions about any new technological design must be answered before it can become a success.

★ **Does the technology meet a need?** The new technology must offer something new or better than existing products, and be something that people truly need.

★ **How will the technology be manufactured?** Henry Ford wanted to sell his new Model T, but he couldn't produce the cars fast enough. He introduced assembly line production so cars could be made more quickly. This new procedure lowered the cost of making cars.

★ **Is the technology easy to operate?** The original clasp-locker failed because it was too complicated. Today, some stores offer classes to people buying a new technology, such as a digital camera, to show them how to use it. User-friendly products are generally adopted more quickly.

★ **Is the technology easy to maintain?** As long as zippers rusted and broke easily, people did not use them.

★ **How will the technology be disposed of?** Society has to think about effects on the environment and other hidden costs before adopting a new technology. For example, people began buying more paints than oil paints when these paints became water-based. Water-based paints clean up easier and can be safely washed in a sink.

★ **How will the technology be sold?** Often new technology is sold in stores, but it can also be sold through the Internet or an independent distributor.

APPLYING WHAT YOU HAVE LEARNED

✦ Identify a new technology and explain what factors led to its adoption. Consider how it is manufactured, operated, maintained, sold and disposed of.

THE RELATIONSHIP BETWEEN SCIENCE AND TECHNOLOGY

Science and technology work together and support each other. Existing technologies help us to meet our needs and solve our problems. Scientific knowledge opens the way for the development of new products. As new technologies develop, they generate new questions that must be answered through scientific inquiry. They also give scientists better tools for research. For example, the invention of telescopes helped scientists learn about the solar system. Science and technology are **interdependent** — each depends on the other.

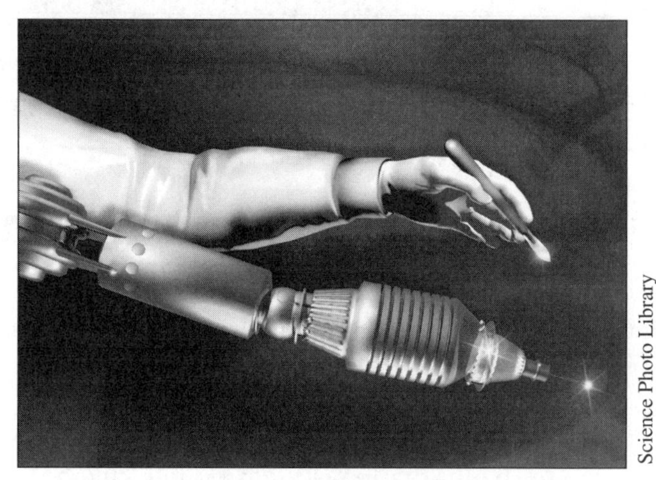

Science Photo Library

Robotic arms are now able to perform surgery

RISKS VS. BENEFITS

The Three Mile Island Nuclear Plant

A new technology can be developed and used only if society determines that its benefits outweigh its risks. Society has to determine what level of risk it can accept in order to gain the expected benefit from a new system or product. It must consider both **short-term** and **long-term** effects. To make this determination, researchers perform a **risk-benefit analysis** that examines all the risks, including side effects and hidden costs, and all the benefits, including the effectiveness and profits of the new technology. For example, some people see more benefits than risks in nuclear power. They argue that it makes us less dependent on foreign countries for oil, and that nuclear power plants do not pollute the air as fossil fuels do. Others say that the risk of a nuclear accident (*like the one at Three Mile Island*), the problem of disposing of nuclear wastes, and the threat of a terrorist attack on a nuclear power plant far outweigh the benefits of using nuclear power.

SCIENCE INFORMS PUBLIC POLICY

Science plays a key role in risk-benefit analysis. Scientific knowledge helps researchers make a more accurate assessment of the risks and benefits of any new technology. Mathematical and statistical models help researchers calculate the risks and benefits based on a small sample, such as when testing new drugs. Before the first nuclear power plant could be built, scientists had to assure lawmakers that it was safe to use to meet society's need for electricity. Before a new drug can be sold, researchers must document that the drug's benefits outweigh its risks. Scientists report facts about new technologies based on their research. Society then interprets these facts to make decisions about what technologies to adopt.

APPLYING WHAT YOU HAVE LEARNED

- Imagine you are on a panel of scientists examining alternative energy sources. List **three** factors you would consider in evaluating different energy sources.

SCIENTIFIC ADVANCES AND TECHNOLOGIES

Scientific advances are occurring faster today than at any other time in history. These scientific advances lay the foundation for rapid technological change. New technologies also affect scientific research. Two examples of such emerging technologies are recombinant DNA technology and nanotechnology. Each has benefits and risks.

CASE STUDY: RECOMBINANT DNA TECHNOLOGY

You will learn about genes and DNA later in this book. Recombinant DNA technology involves transplanting a piece of a gene from one organism to another organism. This changes the genetic makeup of the recipient organism and its offspring, creating a new kind of organism. Through recombinant DNA technology, scientists have created insulin-producing bacteria. These bacteria produce insulin that is needed by people with diabetes.

- **Benefits:** New medicines; fixing genes to cure diseases; new foods; and microbes that can eat wastes or spilled oil.
- **Risks:** Genetic screening could penalize people with hidden diseases; unseen consequences of changing genes, and ethical problems with cloning.

CASE STUDY: NANOTECHNOLOGY

Nanotechnology involves building devices that are molecular in size. Scientists can manipulate one atom at a time, allowing technologists to build devices smaller than ever at lower cost. New materials called nanotubes can conduct heat more efficiently than any other known material. They are stronger than steel and more durable than diamonds.

- **Benefits:** Smaller and more powerful computers; frictionless machines.
- **Risks:** Invasions of privacy by tiny sensors and transmitters; increased military and terrorist threats through the development of new weapons; and the inability of society to keep up with possibly vast changes.

NEW CAREERS IN SCIENCE AND TECHNOLOGY

Rapid advances in science and technology have opened up many new career opportunities, even as they eliminate older jobs. People can find exciting new careers in a variety of science-related occupations. These occupations include:

- marine biologist
- botanist
- agronomist
- climatologist
- ecologist
- entomologist
- astronomer
- geneticist

APPLYING WHAT YOU HAVE LEARNED

- Identify any recent scientific advance or emerging technology you have read about and describe its possible risks and benefits to society.

WHAT YOU SHOULD KNOW

★ You should know that good technological design responds to the needs of society. Many factors have to be considered before a new technology can be adopted.

★ You should know that technology and science influence each other.

★ You should know that scientific advances and emerging technologies have a great impact on society.

CHAPTER STUDY CARDS

Technological Designs

★ **Technology** applies science to meet human needs.

★ **Steps in Designing Technology:**

- **Identify a Problem or Need**
- **Research the Problem**
- **Design a Product**
- **Build a Sample**
- **Test and Evaluate the Design**
- **Adapt and Refine the Design**

Interdependence of Science and Technology

★ **Science and Technology.** Science and technology are interdependent and affect each other.

★ **New Technologies.** New technologies are adopted only after society weighs their risks and benefits. Scientists assist in this analysis.

★ **Scientific Advances.** Rapid scientific advances contribute to emerging technologies and new career opportunities, such as in biotechnology and nanotechnology.

CHECKING YOUR UNDERSTANDING

1 In what way is the development of a new technological design similar to the process of scientific experimentation?

◆ Examine the Question
◆ Recall What You Know
◆ Apply What You Know

A. Most technological designs are developed to help people understand the natural world.
B. Technological designers must consider ease of use by consumers.
C. Technological designers frequently test and revise their designs.
D. Technological designs must be capable of mass production.

ST: A 9–3

This question asks you to compare the processes of technological design and scientific experimentation and tell what they have in common. Recall what you know about each process. Technological designers identify a need, research the problem, design a product, build a sample, and usually test the sample and revise their design based on the results. Scientists ask scientific questions, develop a hypothesis, and test the hypothesis through experimentation. They frequently test and revise their hypothesis and experimental design. Now apply what you know to the question. Answer choice C correctly identifies an important similarity between the two processes.

Now try answering other questions about the role of science and technology.

2 For a new technology to be successful, certain factors must be considered regarding whether or not to adopt the new technology. Identify two factors that scientists would likely consider when trying to decide whether or not to adopt a new technology. (*2 points*)

ST: A 10–3

3 The fields of science and technology are closely related. Many of the modern technological wonders that we use in everyday life are based on scientific work. Identify two examples in which science has influenced modern technology. (*2 points*)

ST: B 10–2

Use the information in the passage below to answer the following question.

ORGAN TRANSPLANTS OF THE FUTURE

Each year, thousands of people need to replace a failing organ with a healthy one. Unfortunately, organs for transplant are in short supply. Moreover, transplants are risky procedures. The recipient's body may recognize the transplanted organ as foreign and reject it. However, scientists are working on taking unspecialized cells called "stem cells" from a patient and trying to grow them in a laboratory. Scientists hope one day to grow these cells into complete organs. Transplants produced by this process would not be foreign and would not be rejected by the immune system of the patient.

4 Organ transplants are an increasingly common part of medical care. Science and technology are working together to make stem-cell transplants a reality. Describe one short-term and one long-term factor that should be considered by scientists in the development of organs grown from stem-cells. (*2 points*)

ST: B 9–1

CHECKLIST OF BENCHMARKS IN THE NATURE OF SCIENCE UNIT

*At the end of each content unit you will find a **Checklist of Benchmarks**. The purpose of these checklists is to help you mentally review the major benchmarks covered in the unit before going on to the next unit.*

Directions. Now that you have completed this unit, place a check mark (✔) next to those benchmarks you understand. If you have trouble recalling information connected with one of the benchmarks, review the chapter indicated in the brackets. For the items you do not recall, reread the section of the chapter dealing with that topic.

SCIENTIFIC WAYS OF KNOWING

- ❑ You should be able to explain how scientific knowledge is based on evidence, predictive, logical, subject to modification and limited to the natural world. **[Chapter 5]**
- ❑ You should be able to explain how scientific inquiry is guided by knowledge, observations, ideas and questions. **[Chapter 5]**
- ❑ You should be able to describe the ethical practices and guidelines in which science operates. **[Chapter 5]**
- ❑ You should be able to recognize that scientific literacy is part of being a knowledgeable citizen. **[Chapter 5]**

SCIENTIFIC INQUIRY

- ❑ You should be able to apply the processes of scientific investigation to create models and to design, conduct, evaluate and communicate the results of these investigations. **[Chapter 6]**

SCIENCE AND TECHNOLOGY

- ❑ You should be able to explain the ways in which the processes of technological design respond to the needs of society. **[Chapter 7]**
- ❑ You should be able to explain how science and technology are interdependent; each driving the other. **[Chapter 7]**

EARTH AND SPACE SCIENCES

In this unit, you will review what you need to know about the Earth and Space Sciences for the **OGT in Science**. You will learn how scientists think our universe may have begun, how stars produce energy, and how our solar system is affected by gravitational forces. Then you will learn about the systems and processes of Earth, including its lithosphere, oceans and atmosphere. Finally, you will study the impact of living organisms on Earth.

N.A.S.A.

As scientists probe deeper into space, they are learning more about the universe

★ **Chapter 8: The Origins of the Universe**

This chapter examines the Space Sciences. By studying the light emitted from stars, scientists have determined that our universe is expanding. Many scientists believe the present universe began with a "big bang" several billion years ago.

★ **Chapter 9: Planet Earth: Systems and Processes**

In recent decades, scientists have learned how tectonic plate movements shape Earth's crust. These movements have shifted continents, caused seafloor spreading, and produced most of the world's volcanoes and earthquakes. Scientists have also discovered many of the key processes of Earth's atmosphere and oceans.

★ **Chapter 10: The Impact of Life on Earth**

This chapter looks at how life forms affect Earth. You will learn how scientists date both rocks and fossils in layers of rock. You will also see some of the ways human activity is affecting Earth's processes.

CHAPTER 8

THE ORIGINS OF THE UNIVERSE

In this chapter, you will learn how scientists are investigating stars and other bodies in outer space to discover the origins of the universe.

MAJOR IDEAS

A. Copernicus proposed the heliocentric theory that planets revolve around the sun.

B. Gravitational forces govern the movements of the planets, comets and asteroids in the solar system.

C. Stars produce energy from nuclear reactions within them.

D. Many scientists believe the universe began with a "big bang" several billion years ago.

EARLY THEORIES ABOUT THE HEAVENS

Astronomy, the study of the stars, planets and outer space, is probably the oldest science. From earliest times, people looked at the day and evening skies and tried to explain what they saw. Ancient peoples charted the stars both for religious purposes and for navigation.

The Greek philosopher **Aristotle** believed that Earth was the center of the universe. He thought that the sun, planets and other stars circled Earth. These ideas were accepted for 2,000 years until the Renaissance, when **Nicolaus Copernicus** proposed that Earth and the other planets actually revolved around the sun. Copernicus based his **heliocentric theory** on the recorded observations of planetary movements. Like many new scientific ideas, the heliocentric theory was at first rejected. It contradicted the teachings of the Catholic Church. However, the new theory was gradually accepted based on evidence from **telescopes**, a new scientific instrument, and other recorded observations.

Library of Congress

Nicolaus Copernicus

GRAVITY AND THE MOVEMENT OF BODIES IN SPACE

Telescopes allowed scientists to observe stars, planets, and comets more closely than ever before. By the late 1600s, scientists were able to explain the movement of planets, planetary moons, and comets through the force of **gravity**. **Sir Isaac Newton** hypothesized that all objects in nature attract one another by this special force. He concluded that the force pulls objects to the ground on Earth also attracts bodies across space towards each other. Gravity decreases quickly as the distance between two objects increases. The strength of this force is proportional to the product of their masses divided by the square of the distance between them:

An original drawing from one of Newton's earlier books on gravity.

Gravity is proportional to: $\dfrac{\text{mass}_1 \times \text{mass}_2}{\text{distance}^2}$

As you can see, if you double the mass of one of the objects, you double the force of gravity. If you double the distance between objects, gravity is reduced to one-fourth.

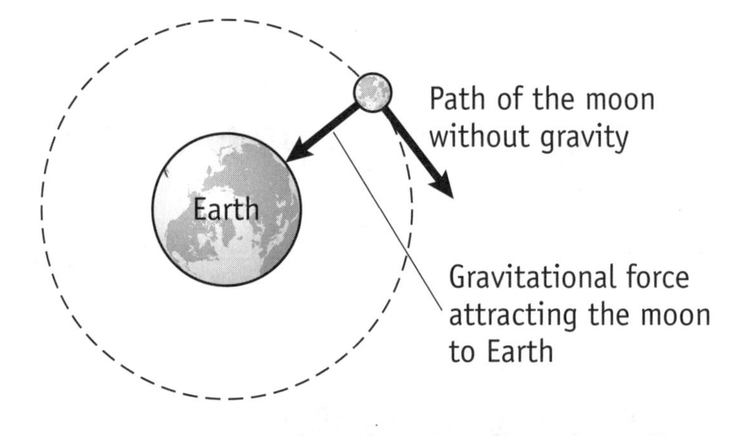

Newton showed that the positions and orbits of objects in space were determined by the laws of motion and gravitational attraction. For example, the moon might travel through space in a straight line, but it is attracted to Earth by the force of gravity. The interaction of these two forces explains why the moon circles Earth.

Newton's discovery showed the relationship between science and mathematics. Scientists now know that the movements of the planets, asteroids, and comets in our solar system are all governed by gravitational forces. Some of the characteristics of these bodies are also affected by gravity. For example, because the moon has less mass than Earth, its gravitational force was not strong enough for it to hold on to its own atmosphere.

N.A.S.A.

The characteristics of the moon are also influenced by gravity

STARS

Today scientists use powerful telescopes to study distant stars. By analyzing the light rays emitted by a star, scientists are able to tell its composition. Most stars are made up mainly of hydrogen and helium. Each star is actually like a nuclear reactor, producing energy through **nuclear fusion** — the joining together of atomic nuclei to make new atoms.

NEBULA

Scientists believe that stars form out of clouds of gases and dust in space known as **nebula**. The force of gravity pulls the gas and dust together. As the cloud becomes more concentrated, it also begins to spin. Small particles come closer and closer together, until they form a star. Tremendous pressure at the center of the star causes single protons (*the center of the hydrogen atom*) to fuse together into helium (*an atom with two protons and two neutrons*). This fusion unleashes tremendous amounts of energy.

Science Photo Library

Optical photograph of a nebula, 900,000 light years across and 500,000 times the mass of the sun

THE COMPOSITION OF STARS

The center of a star is extremely hot and dense. Energy moves from the center through the layers of the star through both **radiation** (*the spread of energy in waves*) and **convection** (*the circular flow of hot, fluid matter*). The energy finally radiates from the surface of the star across space at the speed of light. Eventually, a star will fuse its hydrogen into helium. Without the pressure of nuclear fusion pushing the star outward, gravity causes the star to contract. The internal temperature of the star then rises, causing it to expand again and become a **red giant**. The star now converts helium into carbon, oxygen, and other elements. In fact, all elements in the universe other than hydrogen and helium have been formed by the nuclear reactions of stars. Larger stars develop cores hot enough to fuse particles into iron atoms. Eventually, the star begins to cool. When fusion stops and a large star collapses into itself, it experiences an explosion known as a **supernova**. All elements in the universe heavier than iron have been formed in supernovas. Eventually, the star cools and becomes a **dwarf** or a dense **black hole**.

APPLYING WHAT YOU HAVE LEARNED

- Describe how stars are formed and how they produce energy.
- Explain how elements beyond hydrogen and helium have been formed.

THE "BIG BANG" THEORY

In the last unit, you learned that scientists develop theories to explain what they see in the natural world. They then test their theories by making observations and conducting experiments. One question that has continually fascinated scientists has been: how was the universe first created? Some people hypothesized that the universe had no beginning or end and was infinite. However, scientists now believe the present universe has a history with a definite beginning. Based on observations of the stars, many scientists now believe that the origins of the present universe can be traced back to a single event.

By examining the **spectra** of stars — the patterns of light waves they emit — scientists can see that stars and whole groups of millions of stars, called **galaxies**, are moving apart. This observation has led most scientists to conclude that the universe is expanding. Scientists then simply turn this process backwards, somewhat like watching a movie in reverse, to estimate just how old the present universe is.

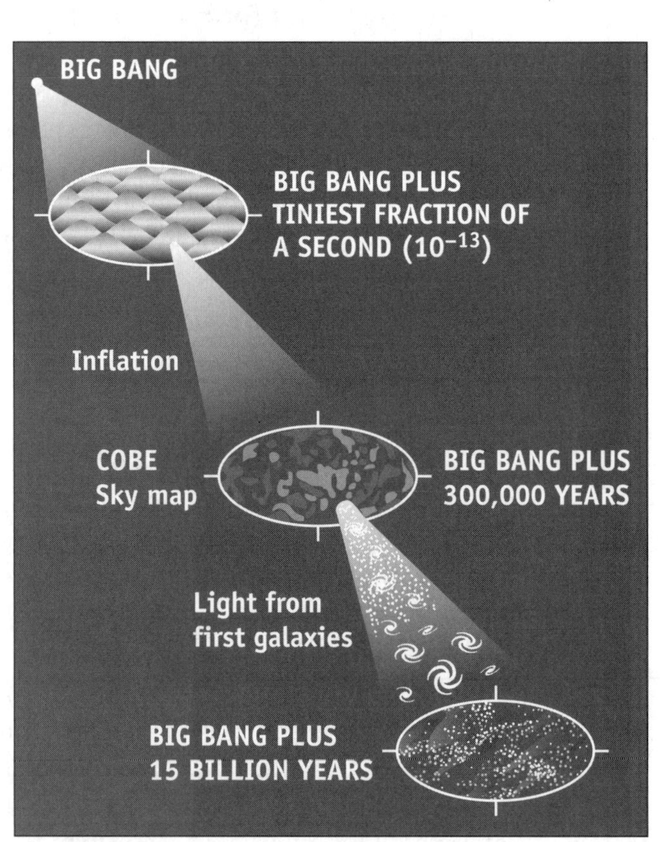

An artist's summary of the "big bang" theory

Given the rate of current expansion, scientists believe that the entire universe began with a single massive explosion that occurred between 13 and 15 billion years ago. According to this theory, it is unknown what existed before the "big bang." Just before this event, all matter and energy in space was contained at one point. Then a tremendous explosion started the universe. Scientists find further support for this theory in the existence of **cosmic background radiation**. This radiation consists of constant microwaves found all over the sky, which many scientists feel is radiation still spreading from the "big bang."

★ In the first billionth of a second after the "big bang," scientists believe the universe was a hot, dense soup of many types of sub-atomic particles (*particles smaller than atoms*), including electrons and pockets of radiant energy called **photons**. At that very instant, the universe was already hundreds of millions of miles wide.

★ After a millionth of a second, the universe had cooled enough for some of these particles to combine as protons and neutrons.

★ One second after the "big bang," scientists believe many of the protons and neutrons collided to form the nuclei of helium atoms (*each with two protons and two neutrons*). After several hundred thousand years, as the universe continued to expand and cool, these hydrogen and helium nuclei captured electrons and formed the first atoms, the building blocks of all matter. You will learn more about atoms in the next unit.

★ Scientists believe that swirling clouds of hydrogen and helium gas now formed. Gravitational forces pulled these clouds together, forming the first stars and galaxies. Within these stars, new chemical elements formed, such as oxygen, silicon and iron.

★ About 10 billion years ago, our galaxy, the ***Milky Way***, began to form. One of the clouds of gas and dust in this galaxy began to form our own solar system. As gravity pulled these particles together, the cloud began to spin into a flat disk with a round center. The center became the sun, while particles in other parts of the disk formed the planets. Earth formed about 4.6 billion years ago. It is believed that ice, dust and particles found in the asteroids and comets of our solar system show what these materials were like before gravity pulled them together into planets. The force of gravitational attraction between the sun and the planets, asteroids, and comets continues to determine their orbits.

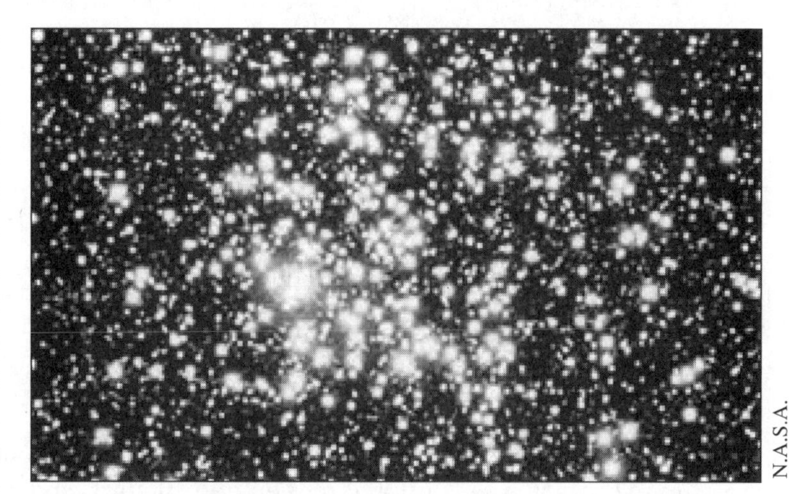
N.A.S.A.

Many scientists believe our universe was created by a massive explosion several billion years ago.

The whole idea of a "big bang" taking place several billion years ago is only a theory that attempts to answer one of the most fundamental questions of all time. No one really knows for sure if this is what really happened. It is important to recognize that this theory is being continually revised based on new evidence. As exploratory telescopes are sent deep into space and further research is conducted, the "big bang" theory becomes more complete and our knowledge of the origins of the universe becomes more accurate.

N.A.S.A.

Probes deep into space help increase our understanding of Earth's origins.

APPLYING WHAT YOU HAVE LEARNED

- Create a timeline of the key events illustrating the "big bang" theory.
- Describe the current scientific evidence in support of the "big bang" theory.

WHAT YOU SHOULD KNOW

- You should know that stars produce massive amounts of energy through nuclear fusion, changing hydrogen atoms (*with one proton*) into helium (*with two protons and two neutrons*), and that some of this energy radiates into space.
- You should know that the force of gravitational attraction influences the movements of the planets, asteroids and comets in our solar system. For example, the gravitational attraction of Earth influences the movement of our moon.
- You should know that many scientists believe that the present universe began with a tremendous explosion, known as the "big bang," between 13 and 15 billion years ago.
- Scientists have formed their theory about the existence of a "big bang" based on evidence from the outward movements of stars, and background cosmic radiation.

CHAPTER STUDY CARDS

The Solar System

- **Heliocentric Theory.** The theory first proposed by Copernicus, stating that Earth and other planets revolve around the sun.
- **Gravity.** First explained Sir Isaac Newton, gravity is a force of attraction between any two objects. Gravity governs the movements and characterization of the planets, asteroids and comets in our solar system. It is proportional to the product of the masses of the object, divided by the square of the distance between them.

Stars and The Big Bang Theory

- **Stars.** Scientists believe stars were formed out of clouds of gases and dust in space known as **nebula.** Stars produce energy through nuclear fusion, converting hydrogen into helium. Stars form all elements besides the lighter gases.
- **"Big Bang" Theory.** Scientists believe that the universe began in an explosion between 13 and 15 billion years ago. They base their theory on the movement of stars away from one another, and from background cosmic radiation.

CHECKING YOUR UNDERSTANDING

1 In each diagram below, the mass of the star is the same. In which diagram is the force of gravity greatest between the star and the planet shown?

- ◆ Examine the Question
- ◆ Recall What You Know
- ◆ Apply What You Know

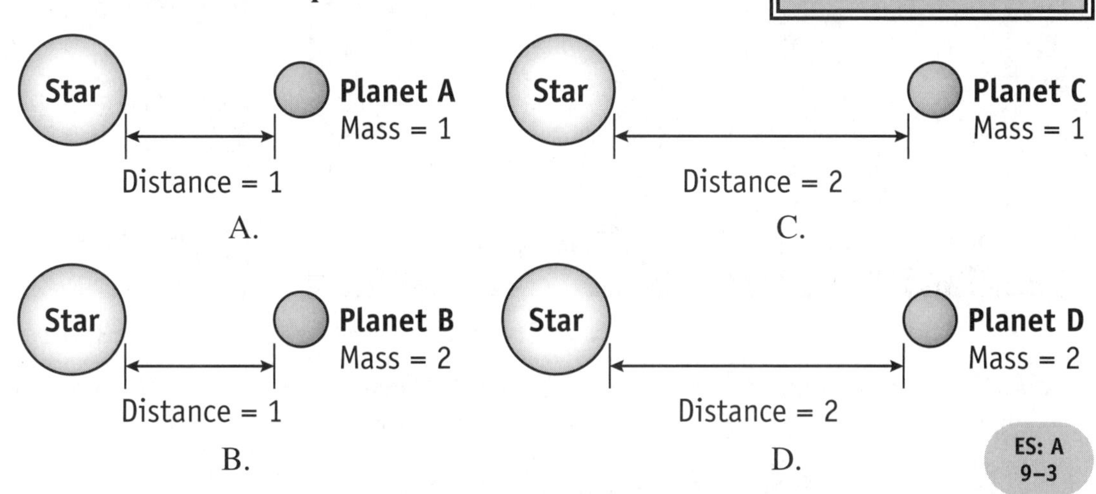

ES: A 9–3

To answer this question, you have to recall that gravity is the force of attraction between any two objects, and that it is proportional to the product of the masses divided by the square of the distance between them: $\left(\frac{mass_1 \times mass_2}{distance^2}\right)$.

In this example, the force of gravity is therefore greatest for answer B.

Now try answering some additional questions on your own about gravity, stars and the universe.

2 Which of the following provides scientific evidence in support of the "big bang" theory?

ES: A 9–2

A. Early Earth was different from the planet we inhabit today.

B. The sun releases energy through nuclear fusion reactions, converting hydrogen to helium.

C. The spectra of stars indicates they are moving apart from one another.

D. The paths of the planets in our solar system are governed by gravitational forces.

3 All elements heavier than hydrogen and helium have been formed by

ES: A 9–1

A. forces below Earth's crust.

B. the "big bang" explosion.

C. processes in stars.

D. supernovae.

Use information from the diagrams below to answer the following question.

Stage 1

A ball of matter and energy exploded.

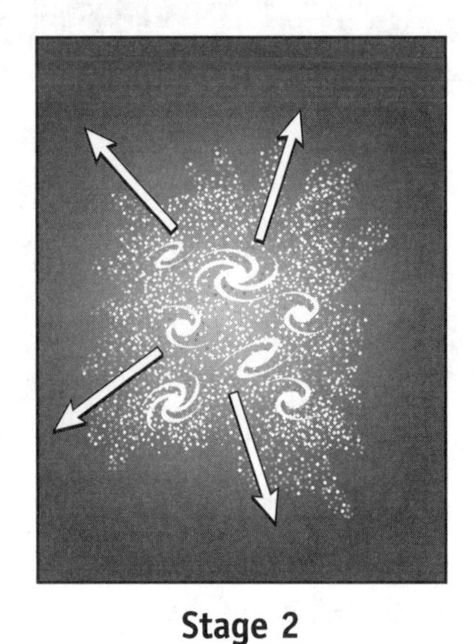

Stage 2

A huge cloud of hydrogen and helium moved outward, condensing to form galaxies.

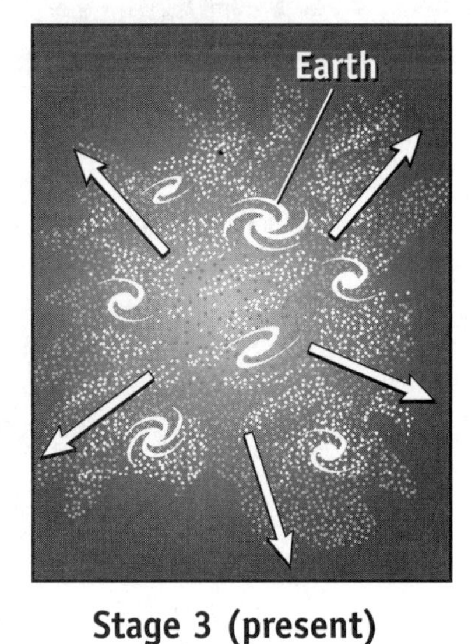

Stage 3 (present)

The galaxies continue to move outward.

4 **The diagram above illustrates a theory popular among scientists to explain the creation of the universe. Identify the theory and provide one piece of evidence scientists use in support of it. (*2 points*)**

ES: A 9–2

5 **If the distance between the moon and Earth were double its present distance, the gravitational force between them would**

A. also double

B. remain the same

C. be one-half of the present force

D. be one-quarter of the present force

◆ Examine the Question
◆ Recall What You Know
◆ Apply What You Know

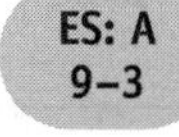

6 **Which graph best represents the relationship between the gravitational attraction of two objects and their distance from each other?**

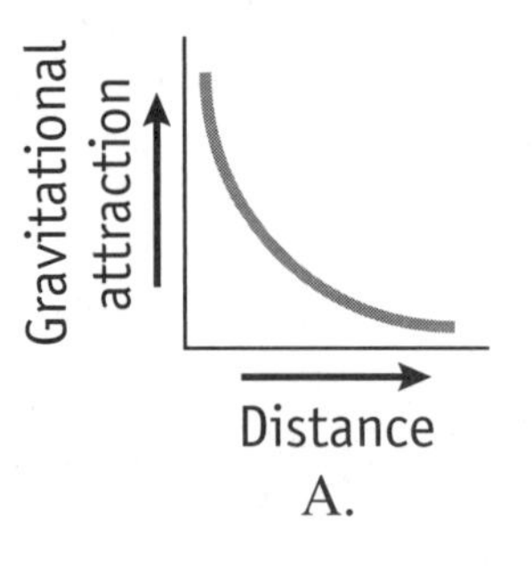

A.

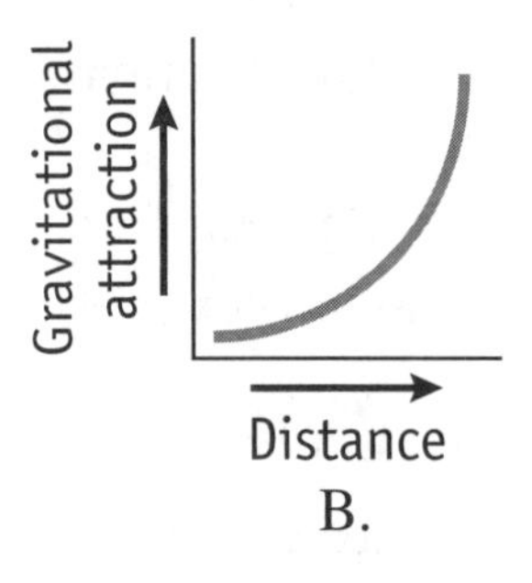

B.

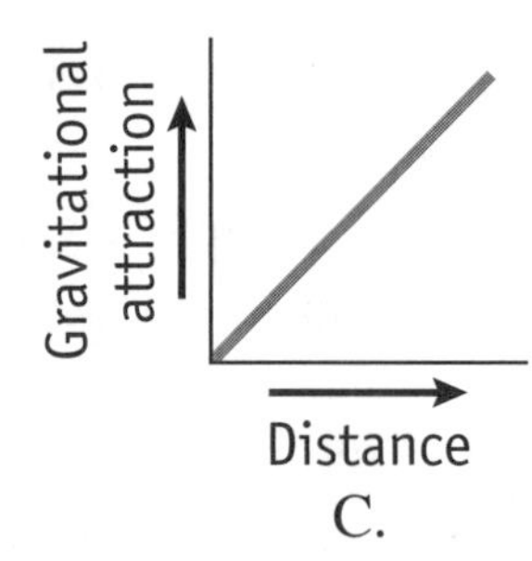

C.

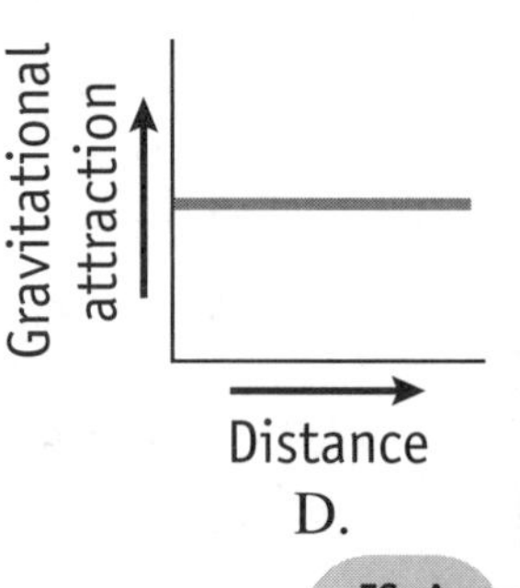

D.

ES: A 9–3

CHAPTER 9

PLANET EARTH: SYSTEMS AND PROCESSES

In this chapter, you will learn about Planet Earth — its structure and processes, including the interior of Earth, its crust, oceans and atmosphere.

MAJOR IDEAS

A. Energy from nuclear radiation and Earth's origins heats the interior of Earth.

B. The movement of tectonic plates shapes Earth's surface, creating mountains, rift valleys, seafloor spreading, earthquakes, volcanoes and tsunamis.

C. Oceans cover most of Earth's surface and help transfer energy through their currents.

D. Water is recycled through evaporation, precipitation and run-off from rivers and groundwater back into the ocean.

E. The atmosphere creates distinct weather patterns and climates at different geographic locations.

THE EARTH'S INTERIOR

By comparing data from **seismic waves** (*waves created by earthquakes*), scientists studying earthquakes and volcanoes have concluded that Earth consists of a series of three distinct layers: **crust**, **mantle** and **core**.

CRUST

The crust forms a thick skin around Earth, much like the crust on a loaf of bread. All life takes place on this topmost layer of Earth. Earth's crust is made of solid rock. **Oceanic crust** is the floor below the oceans. About 5 to 8 km (*kilometers*) thick, it is made of heavy, dense rock. **Continental crust**, in land areas, is much thicker (*on average between 30 and 50 km deep*) than oceanic crust, but it is made of lighter, less dense rock. Continental crust is thickest below high mountains.

MANTLE

Below the crust is an area of hot, dense rock known as the **mantle.** Almost 3,000 km thick, the mantle makes up most of Earth's volume. The top of the mantle is solid, like the crust. As one goes deeper into Earth, both temperature and pressure rise. About 100 km below Earth's surface, the rock is near the melting point and becomes semi-solid or plastic.

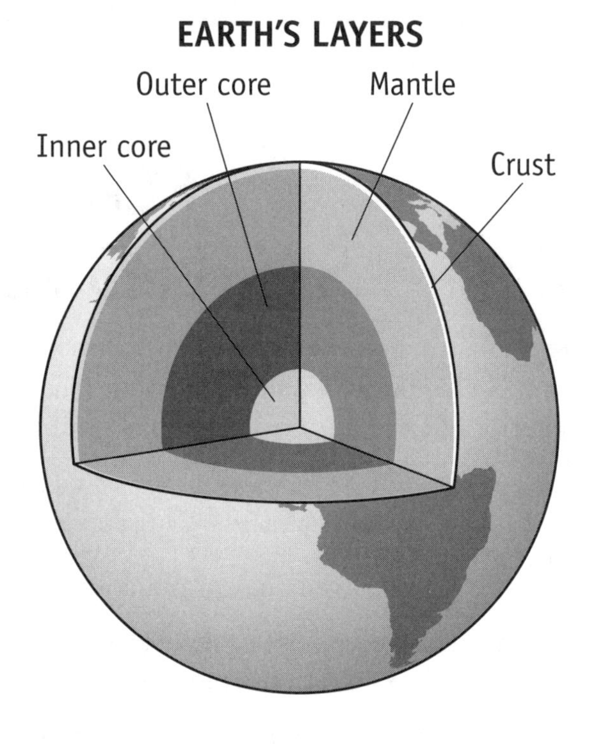

CORE

The center of Earth is known as the **core**. It consists of an inner core of metal, mainly iron and nickel. Earth's core is extremely hot, with temperatures reaching well above 5,000°C. Radiation from radioactive substances and heat energy from Earth's formation create this heat (*geothermal energy*). The outer core is liquid. On the surface of Earth, the metals of the core would boil at these high temperatures, but pressure keeps the outer core in a liquid state. The inner core is made up mainly of iron. Although it is even hotter than the outer core, the tremendous pressure makes the inner core solid. The movement of Earth's metallic core is the source of Earth's magnetic field.

THE LITHOSPHERE: PLATE TECTONICS

In the early 20th century, **Alfred Wegener**, a German scientist, noticed the different continents of Earth seemed to fit together like a giant puzzle. For example, the eastern bulge of South America seemed to fill the space below West Africa. Coal seams and sedimentary rock formations on one continent matched those on another. Mountain ranges that ended at one coastline seemed to begin again on another coastline.

Wegener hypothesized that all continents of the world once fit together into a single, giant continent. Gradually, it separated and its pieces drifted apart. Wegener called this process "**continental drift**." He found further evidence for his theory from the fact that the same fossils were often found in distant places, like West Africa and South America. Fossil remains on both sides of the South Atlantic were from animals known not to be great swimmers.

Many scientists at first rejected Wegener's ideas because he was unable to explain what forces could be powerful enough to move continents. Later evidence indicated Wegener was right. Scientists have found mountain ranges in the Atlantic Ocean and deep trenches in the Pacific Ocean that confirm his theory.

PLATE TECTONIC THEORY

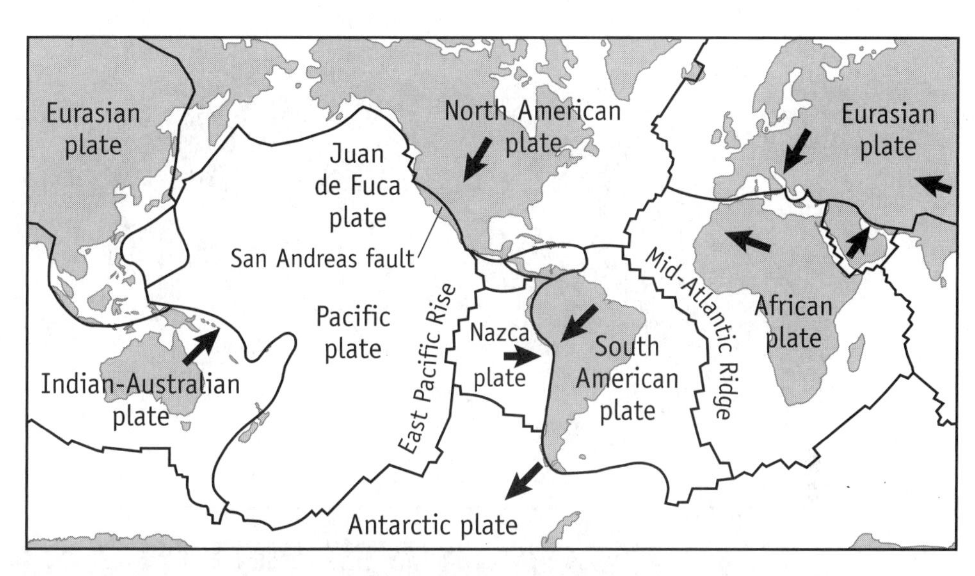

Today, scientists refer to these ideas as the **"plate tectonic" theory**. They identify Earth's crust and part of the solid upper mantle as the **lithosphere**. About 100 km (*60 miles*) thick, the lithosphere is divided into several large slabs of solid rock known as **tectonic plates**. Earth's continents are attached to these plates, although the same plate often includes both oceanic and continental crust. Scientists believe that the tectonic plates act like solid chunks floating on top of the more plastic part of the mantle (*the **asthenosphere***). The plates move at speeds of 1 to 16 cm each year. Over hundreds of millions of years, these plates can move thousands of kilometers.

WHAT CAUSES PLATE MOVEMENT?

Scientists hypothesize that two forces may be responsible for the movement of the tectonic plates:

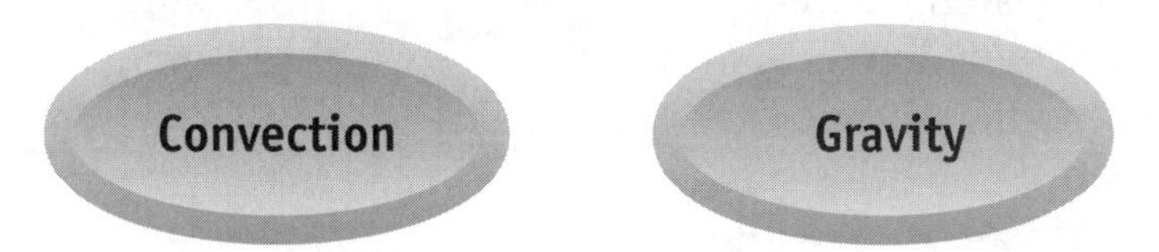

Convection is the spread of heat through the movement of a fluid substance. Inside the mantle, semi-solid rock is heated. As it is heated, it expands and becomes less dense. This lighter rock rises as gravity pulls down cooler, denser rock in its place. After the hotter rock rises and spreads, it begins to cool down. Once cooled, it sinks, creating a circular motion known as a convection current. Scientists believe it may be this **convection current** that pushes the plates above.

Gravity. You already know that the force of gravity is greater when an object's mass is greater. When oceanic and continental plates collide, the denser oceanic plate is pulled by gravity under the lighter continental plate. This process is known **subduction**. As the end of the oceanic plate sinks, it pulls on the rest of the plate as well.

DIAGRAM OF PLATE TECTONICS

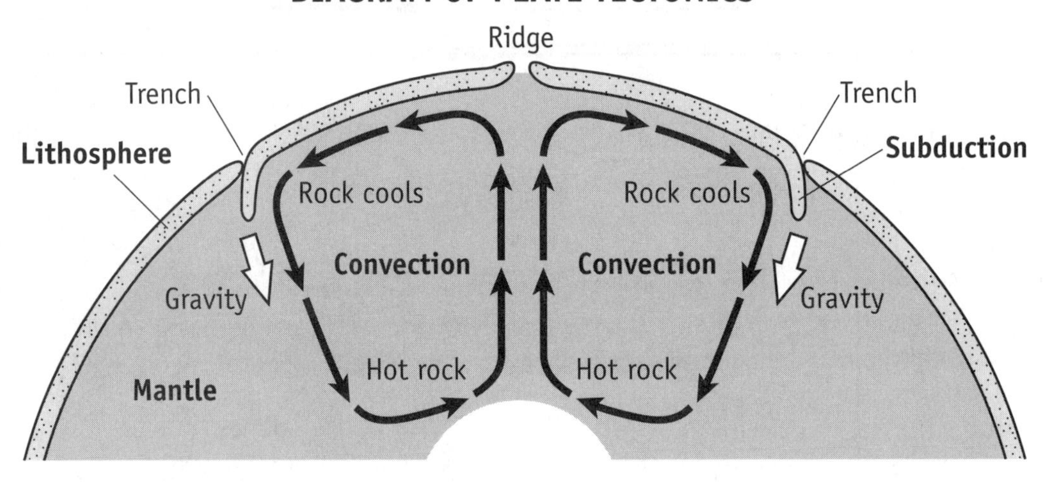

TYPES OF PLATE BOUNDARIES

Tectonic plates push and pull against each other like bumper cars in an amusement park. This results in a variety of different plate boundaries (*places where plates meet*).

DIVERGENT PLATE BOUNDARIES.

Divergent plate boundaries occur where two plates are moving apart, leaving a gap. Hot molten rock, known as **magma**, fills the gap of the plates moving apart, creating a mountainous ridge. As the plates move further apart, a **rift valley** is created between two ridges. There is an important divergent plate boundary in the middle of the Atlantic Ocean, which is adding new crust to Earth's surface. This is known as the **Mid-Atlantic Ridge**.

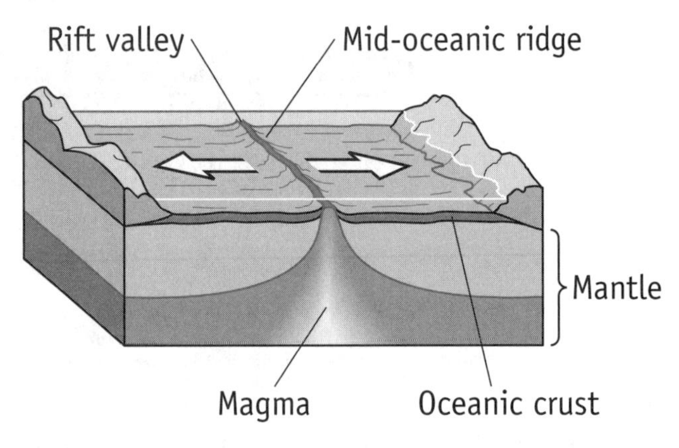

CONVERGENT PLATE BOUNDARIES

Convergent plate boundaries occur where two plates move together in slow collision. This leads to either folding or subduction. For example, the collision of the Indian tectonic plate with the Eurasian tectonic plate has led to the **folding** of Earth's crust and the creation of the Himalaya Mountains and Plateau of Tibet.

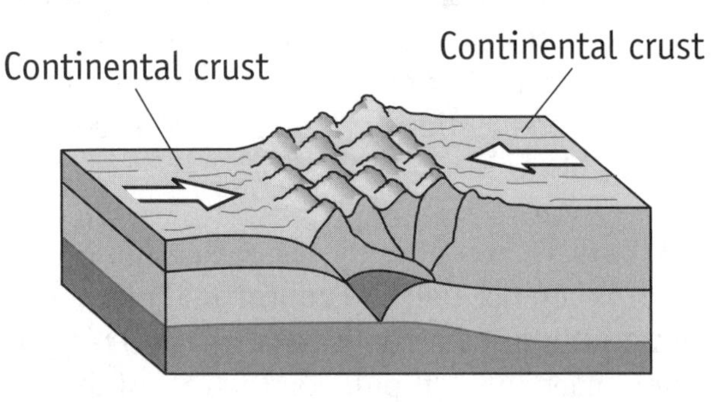

SUBDUCTION ZONE

A **subduction zone** is a type of convergent plate boundary that occurs where a dense oceanic plate collides with a lighter continental plate and into the mantle. Gravity pulls the heavier oceanic plate under the continental plate and into the mantle. This process is occurring around the borders of the Pacific Ocean. Subduction can also lift up the continental plate. For example, the oceanic plate slipping under the western side of South America has lifted up the continental plate, creating the Andes Mountains.

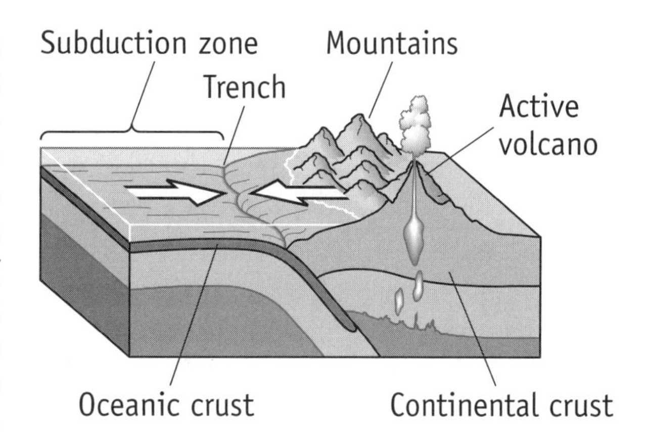

TRANSFORM PLATE BOUNDARIES

A **transform plate boundary** occurs where one tectonic plate slides by another plate horizontally. The San Andreas fault in California is an example of a transform plate boundary, where the Pacific plate slides past the North American plate.

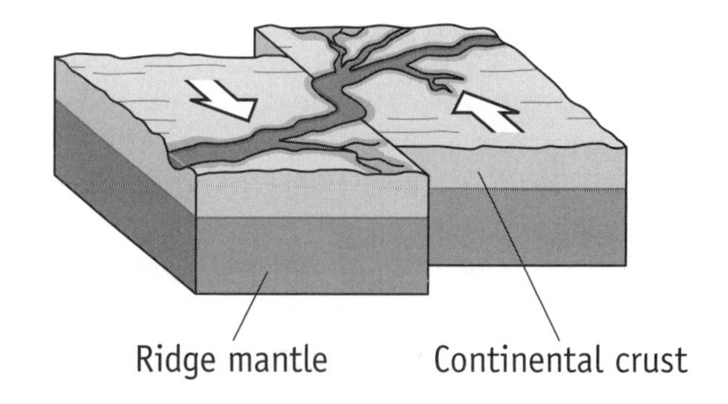

APPLYING WHAT YOU HAVE LEARNED

- ✦ Explain how gravity and convection currents lead to the slow movement of material within Earth.
- ✦ Make a table that identifies and describes the four types of plate boundaries.

EFFECTS OF TECTONIC PLATE MOVEMENT

The movements of tectonic plates explain many of Earth's surface features:

SEAFLOOR SPREADING

In the mid-Atlantic, measurements taken by scientists show that the separation of plates is actually causing the seafloor to spread. As the plates move apart, magma rises through the cracks of the sea floor creating a ridge of mountains. This creation of new crust would increase Earth's size, except that it is balanced by the destruction of crust through folding and subduction.

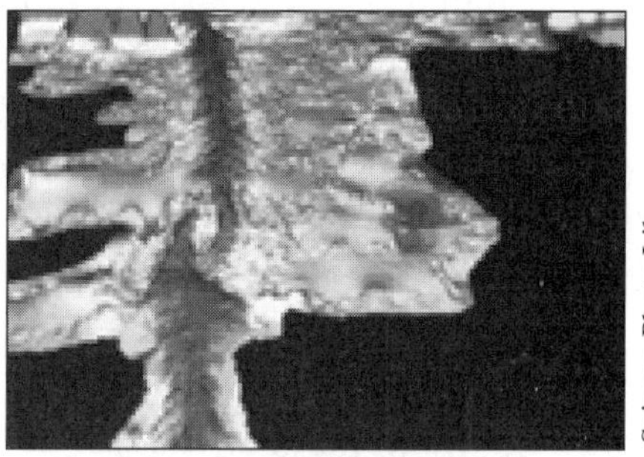

Science Photo Library

Topographic map of a part of the Mid-Atlantic Ridge

MAGNETIC STRIPING

When molten rock fills the gap between divergent plates, iron minerals in the rock line up with Earth's magnetic field. Earth's poles reverse themselves every several hundred thousand years — so that north becomes south. The alignment of iron minerals in the magma reverses itself, too. As a result, scientists have found alternating bands in which magnetic fields are reversed all along the Mid-Atlantic Ridge, providing additional evidence for tectonic plate theory.

FOLDING

When continental plates collide, they cause a folding of Earth's crust that creates new mountains. Such folding has led to the formation of the Himalaya Mountains and Plateau of Tibet — the highest landforms on Earth.

EARTHQUAKES

Plate movements can cause a break in Earth's crust, known as a **fault**. Plate movements can also cause vibrations known as **earthquakes**. As plates move, they create tremendous stress at plate boundaries. Eventually, parts of the rocky crust break, creating a fault and sending vibrations known as **seismic waves**. Scientists measure the waves sent by an earthquake with a **seismograph**. They can see that most waves originate at plate boundaries. The **Richter Scale** is used to measure the amount of energy released by an earthquake on a scale of 2 to 10. When an earthquake occurs under or near the ocean, it creates immense ocean waves of destructive force known as **tsunamis**. In December 2004, an earthquake in the Indian Ocean caused a devastating tsunami affecting South Asia and East Africa.

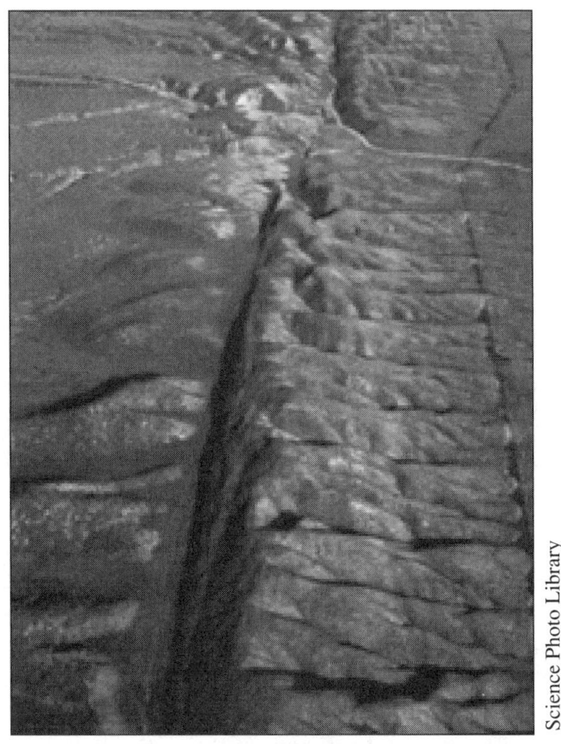

Science Photo Library

View of the San Andreas fault in Central California

VOLCANOES

In places where tectonic plates diverge or where one plate dives under another, pressure in Earth's mantle is reduced and some of the hot, solid rock turns to liquid. Any part of the tectonic plate that sinks into the mantle may also melt. Pockets of molten rock form beneath Earth's surface. This magma may break through weaknesses in Earth's crust. Magma, ashes and gases erupt and form a **volcano**. Once the magma reaches Earth's surface, it becomes known as **lava**. The location of most volcanoes and earthquakes has been shown to be almost identical with the location of plate boundaries. Often volcanoes appear on continental crust that is sitting over subducted oceanic crust. For example, the "Ring of Fire" around the Pacific Ocean — a zone of volcanoes and frequent earthquakes — coincides with the boundaries of the Pacific tectonic plate.

THE CYCLE OF ROCK FORMATION

The rocks of Earth's crust go through cycles influenced by the movements of tectonic plates and other factors. Cooled magma forms **igneous rock**, such as granite or basalt. Erosion from water and air breaks down rocks on Earth's surface into pebbles, sand and dust. These fragments pile up and become compressed as **sedimentary rock**, like sandstone. Changes from tectonic plate movements may push sedimentary and igneous rocks below Earth's surface. Heat and pressure can change these rocks into **metamorphic rock**, such as marble or slate, or even melt the rock completely, so that it forms new igneous rock.

APPLYING WHAT YOU HAVE LEARNED

- ✦ Explain how the actions of tectonic plates have helped shape Earth's surface.
- ✦ Identify two types of evidence that support plate tectonic theory. For each type, explain how that evidence helps to support the existence of tectonic plates.

THE HYDROSPHERE: EARTH'S OCEANS

More than 70% of Earth's surface is covered by water. Scientists refer to this as the **hydrosphere**. About 97 percent of this water is in the oceans; the rest is either frozen in the polar ice caps or found in Earth's atmosphere, groundwater, and freshwater lakes and rivers.

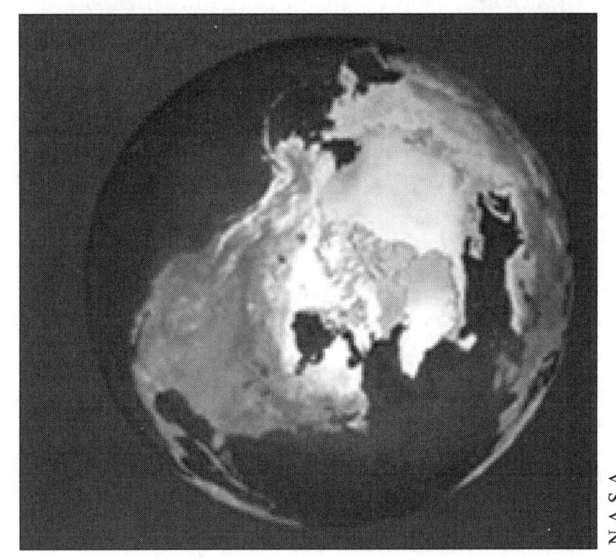
N.A.S.A.

Earth's oceans hold most of its water

THE WATER CYCLE

Just as Earth's lithosphere undergoes various processes, so does the hydrosphere. One of the most important is the water cycle. Solar energy heats the surface of the oceans, causing some of the surface water to evaporate into the atmosphere. Plants also create water vapor through **transpiration**. Water vapor rises until it becomes cooler, then condenses into tiny droplets small enough to float in the atmosphere as clouds. When the droplets grow heavier, they fall back to Earth's surface as **precipitation** — rain, snow or hail. Some precipitation returns to the ocean, but some falls on land where it is absorbed by the ground or forms lakes, streams and rivers. Some precipitation evaporates, but much of the groundwater and rivers eventually drain back into the oceans.

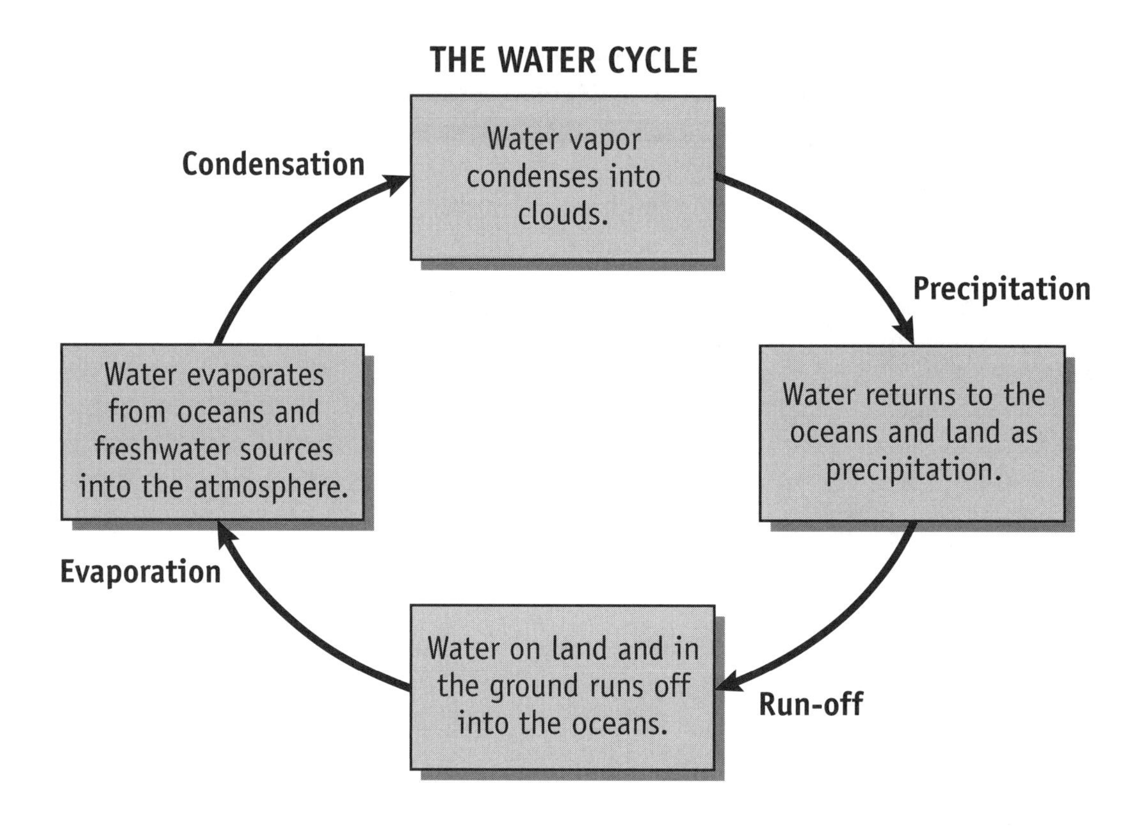

TIDES, CURRENTS AND WAVES

Earth's ocean waters are in constant motion. This can be seen in the tides, currents and waves.

TIDES

Each day, the surface level of the oceans rises up and falls down during ***high*** and ***low tide***. Tides are caused by the gravitational pull of the moon and sun on Earth's ocean waters. The ocean directly facing the moon bulges towards the moon, creating high tide, a time when sea levels are at their highest. On the opposite side of Earth is another high tide, caused by the force of Earth's spin where the moon's pull is weakest. Sea levels become highest when the moon and sun are both lined up on the same side of Earth, and lowest when they are on opposite sides. The tides are a remarkable demonstration of the effects of gravity.

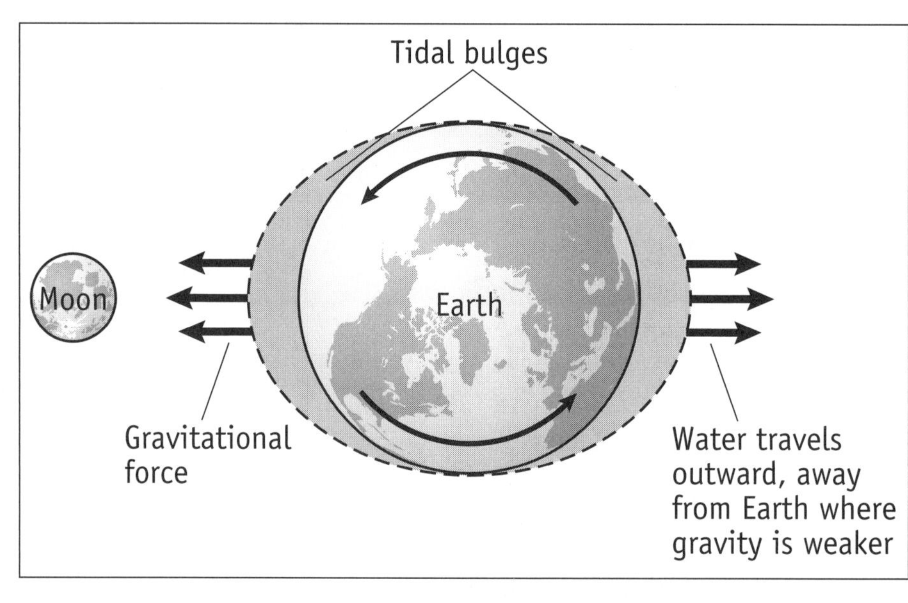

CURRENTS

Ocean currents are movements of the ocean's waters. These currents occur both at the surface and below. Surface currents are mainly caused by the spinning of Earth, and winds. Deeper ocean currents are caused by differences in the temperature and **salinity** (*saltiness*) of ocean water.

At the equator, the spinning of Earth and winds push surface water towards the west. This sets in motion large circular surface currents. Water heated by the sun moves away from the equator, carrying heat energy towards the polar regions. For example, the Gulf Stream carries warm water towards Great Britain, making that country warmer than it would otherwise be. This transfer of energy helps maintain a balance — carrying heat from the tropics to colder regions, and cold from the polar regions towards the tropics.

Below the ocean's surface, its waters actually separate into different layers based on their density. Cold, salty water is more dense than warm, less salty water. At the poles, cold, salty water sinks. It then slowly moves towards the equator, pushing warmer water away. During this process, this cold water gradually warms up as it absorbs heat from the layers of water above it. This slow but steady circulation of the ocean's deep waters takes hundreds of years.

WAVES

Ocean waves appear to carry water, but in fact they only transfer energy. Water moves up and down on the ocean's surface in waves but does not actually move across with the wave. Winds, tides, and earthquakes all cause waves.

INTERACTION OF THE OCEANS AND LAND FORMS

The world's oceans interact with its land forms in a variety of ways.

Rivers carry sediment, including salts, into the ocean. Most of the ocean floor is covered by this sediment, which has taken millions of years to accumulate. Ocean currents carry some of this sediment to coastlines, where it forms sandy beaches. The actions of tides and waves can also erode shorelines, chipping away at rock and dissolving minerals.

N.A.S.A.

Ocean currents help explain the reason for some landform shapes.

THE ATMOSPHERE AND CLIMATE

Around Earth is an envelope of gases known as the **atmosphere**, consisting mainly of nitrogen (78%) and oxygen (21%). The atmosphere absorbs solar radiation, moderates temperatures, and distributes water. The atmosphere consists of several layers, but the weather on Earth's surface (*precipitation, winds, and temperature*) is mainly determined by the **troposphere**, the layer closest to Earth. Different processes in the atmosphere lead to differences in **climate**, the average weather conditions occurring in a place over a long period of time.

WEATHER PATTERNS AND LOCATION

Weather patterns and climate are often the result of location or special geographic features. For example, temperatures are generally warmer the closer one gets to the equator. Temperatures become cooler at higher altitudes, such as on mountains or high plateaus. Because air is cooled as it rises above a mountain, the ocean side of a mountain often has heavy precipitation. The air loses moisture and becomes drier as it reaches the other side of the mountain, which may have very little rain or snow.

TORNADO ALLEYS

The spinning of Earth and the atmosphere's uneven heating create specific **wind** patterns. Cold sinking air creates areas of high pressure, while hot rising air creates areas of low pressure. Winds then blow from the high to low pressure areas. **Tornadoes** are high-speed winds that whirl in a funnel. They generally occur in the Great Plains of the Central United States. A tornado occurs when dry, cool air meets warm, humid air. Warm air at the center of tornadoes rises quickly, sucking in both air and objects.

TROPICAL HURRICANES

Hurricanes occur in tropical regions in late summer and early fall when the ocean water is very warm. The warm ocean water evaporates so quickly that it creates an area of low pressure. Air around the rising air column begins to spiral at high speeds. The hot air rises until it cools and condenses, releasing energy, causing heavy rains, winds and lightning.

National Oceanic and Atmospheric Adm.

A hurricane approaches landfall on the coast of Mexico

"LAKE EFFECT" SNOW

Water requires energy to change from a solid (*ice*) to a liquid, or from a liquid to gaseous water vapor. Its temperature therefore changes more slowly than land. Oceans and lakes are therefore cooler than neighboring land areas in summer and warmer than neighboring land areas in winter. One effect of this is "lake effect" snow. Areas near lakes often have heavy snow in winter. This is because large lakes are slow to freeze. As winds blow across the lake, they pick up water vapor. The water vapor condenses and becomes snow when the air passes over colder land areas. For example, winds blowing from west to east across the Great Lakes bring heavy snow to Buffalo, New York.

APPLYING WHAT YOU HAVE LEARNED

- ✦ The interaction of Earth's atmosphere and oceans results in several characteristic patterns on Earth's surface. Identify two ways in which the world's oceans and atmosphere interact and describe their effects.

CLIMATE AND BIOMES

The **biosphere** refers to all life on Earth. Weather patterns and climate influence what kinds of life forms can successfully live in a particular geographic location. Based on the interaction of climate, natural resources and life forms, scientists have identified several different **biomes**, or geographic regions with particular types of plant and animal life.

TEMPERATE DECIDUOUS FOREST

Temperate deciduous forests develop in mid-latitude regions where there is ample rain and moderate temperatures with cool winters. Trees change colors in fall and lose their leaves in winter. There is a wide range of plant and animal life.

TROPICAL RAIN FORESTS

Tropical rain forests develop in areas near the equator where there is ample rainfall and warm temperatures year-round. Large trees cover these areas with their leaves, forming a **canopy**. Despite the rapid growth of trees, the topsoil is actually very thin. Tropical rain forests are marked by a great abundance of animal and plant life, enjoying greater biological diversity (*known as biodiversity*) than any other biome.

GRASSLANDS

Grassland areas exist where the climate is drier and there is not enough rainfall to support large amounts of trees. Instead, grasses dominate, as well as large grazing animals, like cattle, antelope or bison.

DESERTS

Deserts are regions that receive less than 10 inches of rainfall annually. Deserts in the tropical latitudes, such as the Sahara Desert, have their own special forms of plant and animal life, which have adapted to the lack of water and extremes of temperature. Cacti, for example, store water in their stems.

TUNDRA

Tundra is found closer to the polar regions where the soil is so cold that trees cannot grow. Much of the ground is frozen part of the year. Tundras constitute a distinct biome, with their own plant and animal life, including grasses, small shrubs. Large mammals and birds migrate to these regions in the warmer spring and summer months.

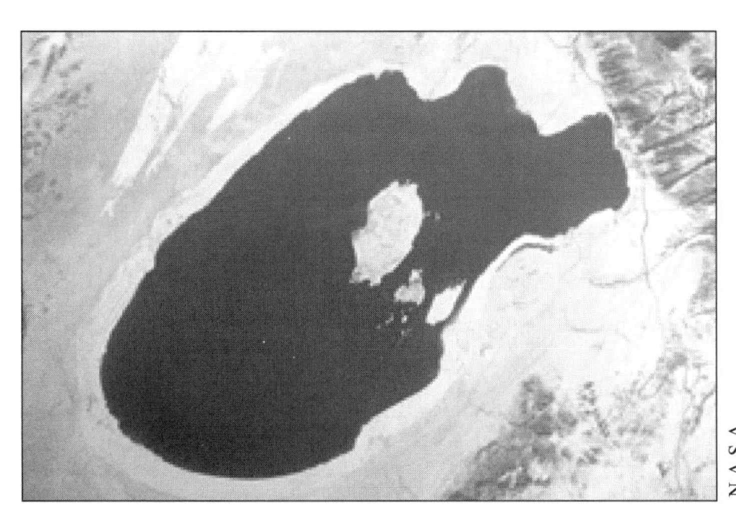
N.A.S.A.

In a tundra, much of the ground is frozen.

WHAT YOU SHOULD KNOW

You should know that many processes occur in patterns in Earth's systems:

★ **The Lithosphere.** The lithosphere is made up of the crust and solid upper mantle, and is broken up into tectonic plates. The movement of these plates moves and shapes Earth's surface. New crust is made when plates spread apart at divergent plate boundaries. Crust folds into new mountain chains where continental plates collide. Ocean crust folds under lighter continental crust at subduction zones. Earthquakes and volcanoes most often occur at plate boundaries.

★ **The Hydrosphere.** The world's oceans cover over 70% of the planet's surface area. Currents, caused by winds and differences in water density, move the ocean's waters around the planet. Water is also moved between the atmosphere and land surfaces through the cycle of evaporation and condensation.

★ **The Atmosphere.** The spinning of Earth, the unequal heating of air by the sun, the evaporation of water, and the effect of various landforms on the air result in weather — variations in temperature, wind, and precipitation. Climatic differences give rise to distinct ***biomes*** — geographic regions supporting different kinds of life, such as forests, grasslands, deserts and tundra.

CHAPTER STUDY CARDS

The Earth's Interior

★ **Crust.** Outermost surface of Earth; oceanic crust is the floor beneath the oceans and is thinner and denser than continental crust. Continental crust in land areas is much thicker than oceanic crust.

★ **Mantle.** Almost 3,000 km thick, the mantle is made up of hot, dense rock. As one moves deeper, the temperature and pressure rise.

★ **Outer Core.** Molten nickel and iron.

★ **Inner Core.** Solid, mainly iron.

Tectonic Plate Movements

★ **Lithosphere.** Crust and top layer of mantle; divided into shifting tectonic plates.

★ **Plate Boundaries**

- **Divergent.** Two tectonic plates spread apart. Magma comes through the gap.
- **Convergent.** Two tectonic plates come together.
- **Subduction Zone.** Oceanic crust dives under lighter continental crust.
- **Transform Plate Boundaries.** Two plates slide by horizontally.

Hydrosphere

★ The hydrosphere is made up of all water on Earth's surface.

★ 70% of Earth's surface is covered by oceans.

★ The gravitational pull of the moon causes tides — cyclical rise and fall of the oceans.

★ Ocean water is moved by surface and deep-sea currents.

★ **The Water Cycle.** Water circulates through evaporation, condensation, precipitation, and run-off.

Atmosphere

★ The atmosphere is an envelope of gases around Earth. It is mainly made up of nitrogen and oxygen.

★ These gases absorb solar radiation, moderate temperatures and distribute water.

★ The atmosphere creates distinct weather patterns. Heating of the atmosphere and Earth's spin create wind patterns. Surface features like mountains also affect weather.

★ Variations in climate lead to different ***biomes*** such as temperate deciduous forests.

APPLYING WHAT YOU HAVE LEARNED

✦ Make two cards on your own with the following topics:

- **Plate Tectonics: Causes and Effects**
- **Weather Patterns and Biomes**

CHECKING YOUR UNDERSTANDING

1 Scientists believe Earth's outer and inner cores are both composed of

A. liquid

B. solid

C. a high percentage of iron

D. the same pressure

♦ Examine the Question
♦ Recall What You Know
♦ Apply What You Know

This question requires you to recall information about Earth's structure and processes. You should recall that the outer core is made up of extremely hot liquid. The inner core, although even hotter than the outer core, is solid due to the extreme pressure. Knowing this information means that choices A, B, and D cannot be correct. Both the outer and inner core, however, are mainly composed of iron. Thus, choice C is the correct answer.

Now try answering some questions on your own about Earth's interior, lithosphere, hydrosphere, and the atmosphere.

2 The rising of hot, semi-solid rock in Earth's mantle and the sinking of cooler rock is known as

A. condensation.
B. radiation.
C. convection.
D. metamorphism.

ES: B 9–5

3 Which geologic events occur most often at a mid-oceanic ridge plate boundary?

A. magnetic pole reversals and cooling of ocean water
B. meteorite impacts and tilting of shorelines
C. hydrospheric pollution and adiabatic heating
D. volcanic eruptions and the creation of new crust

- Examine the Question
- Recall What You Know
- Apply What You Know

ES: E 9–6

4 Rocks are classified as igneous, sedimentary, or metamorphic based primarily on their

A. texture.
B. crystal or grain size.
C. method of formation.
D. mineral composition.

ES: E 9–6

5 When a continental crustal plate collides with an oceanic crustal plate, the continental crust is forced to move over the oceanic crust. What is the primary reason that the continental crust stays on top of the oceanic crust?

A. Continental crust is less dense.
B. Continental crust deforms less easily.
C. Continental crust melts at higher temperatures.
D. Continental crust contains more minerals.

ES: E 9–5

6 Which biome is correctly paired with its climatic zone?

A. moderate rainfall and temperatures — deciduous forest
B. heavy rainfall and warm temperatures — tundra
C. little rainfall and warm temperatures — tropical rainforest
D. moderate rainfall and cold temperatures — desert

ES: B 10–1

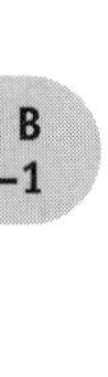

Use the information in the passage to answer the following question:

In the 1930s, most scientists believed that Earth's crust and interior were solid and motionless. A small group of scientists talked about "continental drift" — the idea that Earth's crust is not stationary, but is constantly shifting. From seismic data, geophysical evidence, and laboratory experiments, scientists now believe that lithospheric plates move at the surface. Both Earth's surface and interior are in motion. Solid rock in the mantle can be softened and shaped when subjected to the heat and pressure within Earth's interior over millions of years. Gravitational forces and convection processes are believed to be the driving force of plate tectonics. This theory cannot be directly observed and confirmed. The details of why and how plates move continues to challenge scientists.

7 In this passage, Earth's crust is described as "constantly moving." Give two examples of evidence that supports the conclusion that continents have drifted apart. (*2 points*)

ES: E 9–7

---

8 The cross section below shows the prevailing winds that cause different climates on the two sides of this mountain range.

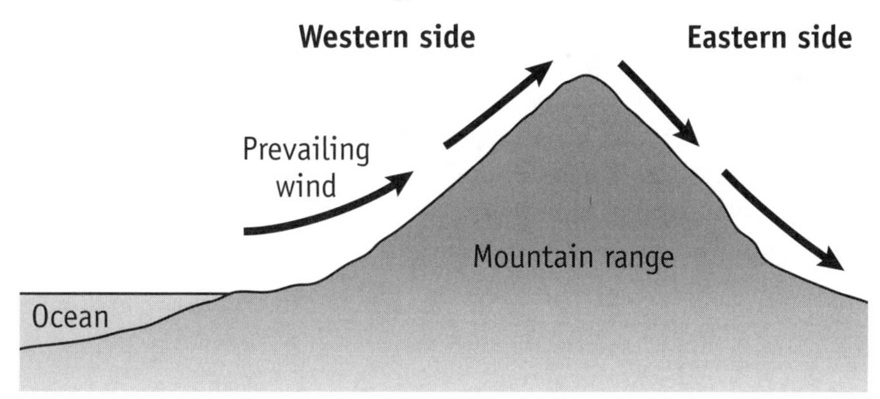

Compared to the climate conditions on the western side of this mountain range, the conditions on the eastern side are more likely to receive

A. frequent hurricanes.
B. larger snow accumulations.
C. heavy rains.
D. less precipitation.

ES: B 10–2

---

9 Clouds form in tropical hurricanes because the air is

A. sinking, expanding and then cooling.
B. sinking, compressing, and then warming.
C. rising, expanding, and then cooling.
D. rising, compressing, and then warming.

- Examine the Question
- Recall What You Know
- Apply What You Know

ES: B 10–2

The map to the right shows the present-day locations of South America and Africa. Remains of *Mesosaurus*, an extinct freshwater reptile, have been found in similarly aged bedrock formed from lake sediments at locations X and Y.

10 Which statement represents the most logical conclusion to draw from this evidence?

A. *Mesosaurus* migrated across the ocean from location X to location Y.

B. *Mesosaurus* came into existence on several widely separated continents at different times.

C. The continents of South America and Africa were joined together when *Mesosaurus* lived.

D. The present climates at locations X and Y are similar.

ES: E
9–7

---

11 Approximately 100 km thick on average, the solid outer layer of Earth is divided into large slabs of solid rock known as tectonic plates. Over long periods of time, these tectonic plates separate, collide, or slide by each other. Identify two effects of plate tectonics. For each effect, explain how plate tectonics causes that effect. (*4 points*)

ES: E
9–6

Use the map below to answer the following question.

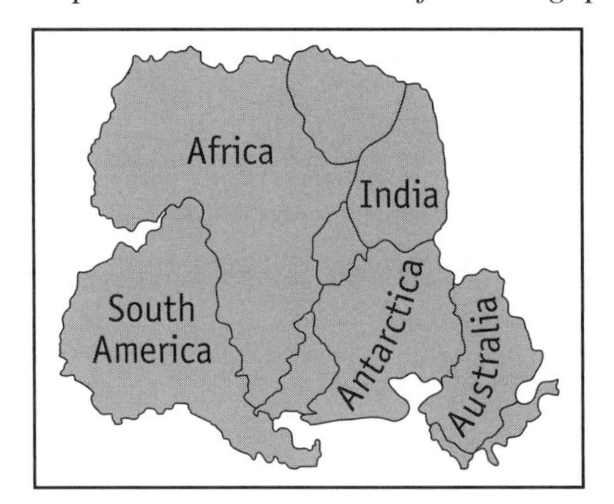

- ♦ Examine the Question
- ♦ Recall What You Know
- ♦ Apply What You Know

12 The diagram above illustrates the scientific theory stating that

A. millions of years ago Earth's land masses formed one giant continent.

B. Earth's continents are slowly moving to become one giant continent.

C. Earth moves around the sun.

D. there is little difference in climatic patterns around the world.

ES: F
9–8

Use the information in the diagram below to answer questions 13 to 15.

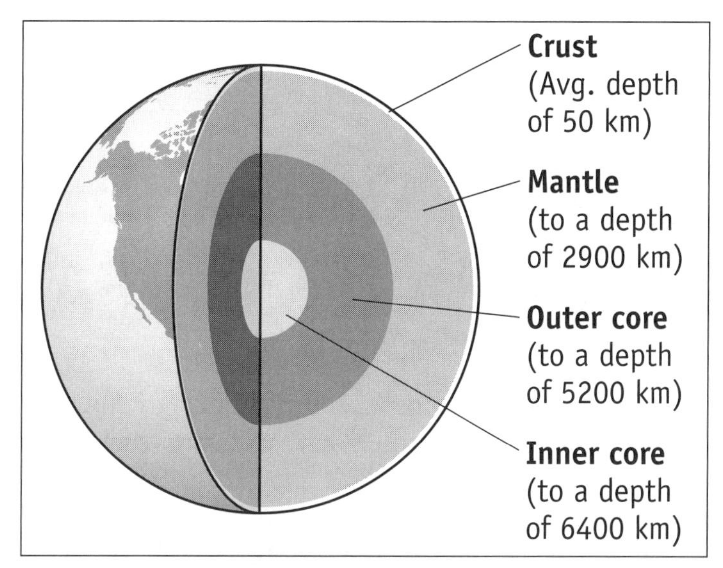

13 The thinnest section of Earth's crust is found beneath Earth's

A. oceans.
B. mountain regions.
C. desert regions.
D. coastal plains.

ES: B 9–5

- ♦ Examine the Question
- ♦ Recall What You Know
- ♦ Apply What You Know

14 In which layer of Earth's interior do scientists believe the average temperature is approximately 5000°C?

A. crust
B. mantle
C. outer core
D. inner core

ES: B 9–5

15 Scientists have classified Earth's interior into the layers shown in the diagram based primarily on evidence gained from

A. drilling for oil.
B. studying volcanic eruptions.
C. measuring seismic activity.
D. measuring cyclone activity.

ES: B 9–5

16 Which of the following graphs accurately represents the relationship between the depth below Earth's surface and density?

ES: B 9–5

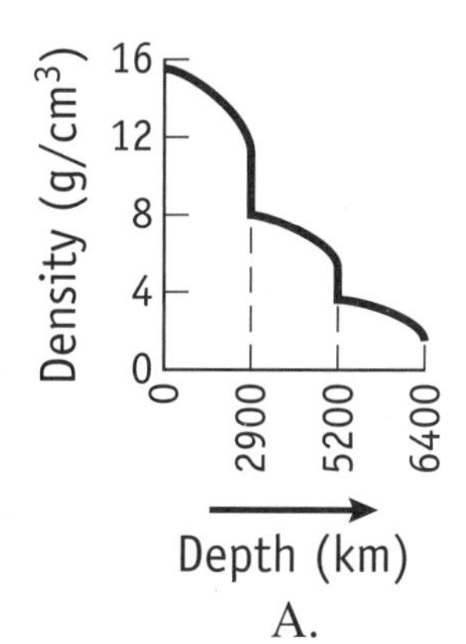

A.

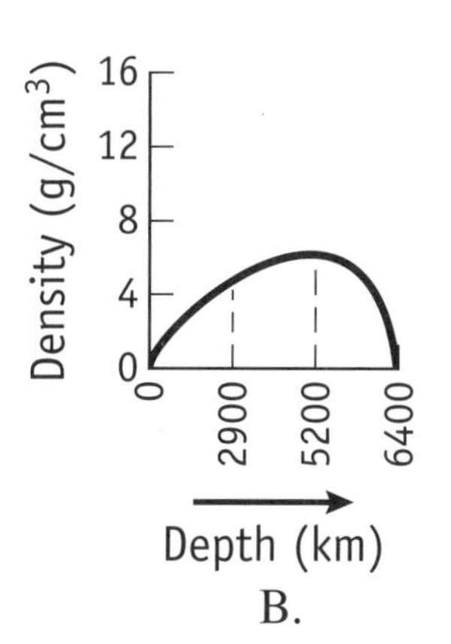

B.

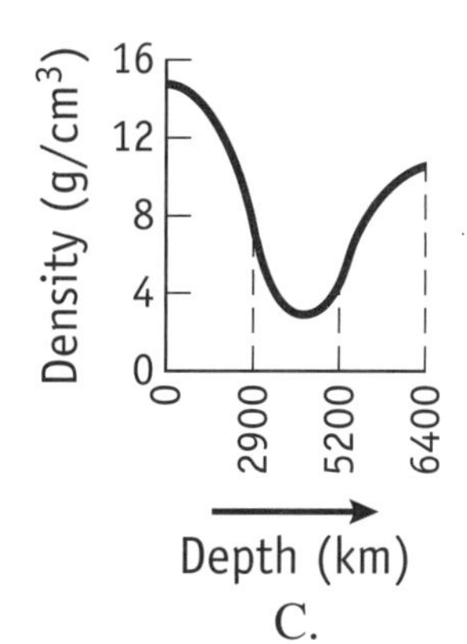

C.

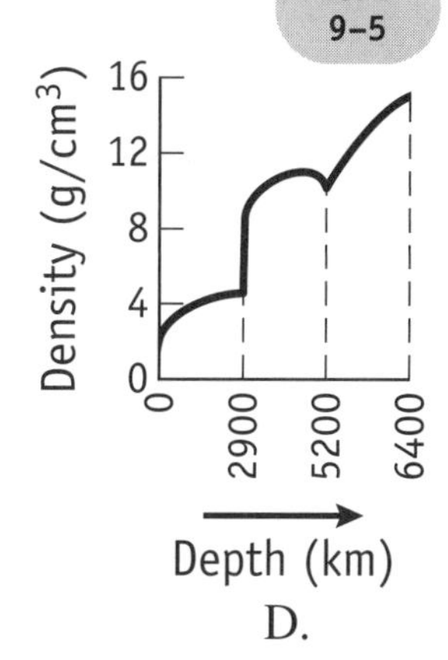

D.

CHAPTER 10

GEOLOGIC TIME AND THE IMPACT OF LIFE ON EARTH

In this chapter, you will learn about the impact of living organisms on the development of Earth's systems. You will also learn how human activities are causing accelerated changes to our planet.

MAJOR IDEAS

A. Scientists are able to estimate the age of rocks and fossils both by studying rock layers and by radiometric dating.

B. Life first appeared on Earth about 4 billion years ago.

C. The processes of living organisms gradually added oxygen to Earth's primitive atmosphere.

D. Human activities are affecting Earth's natural processes and using up natural resources.

DATING ROCKS AND GEOLOGIC TIME

People once believed that Earth was only several thousand years old. Geologists believed that a series of unique catastrophes or supernatural events had helped to shape Earth's surface. However, in 1830 an English geologist, Sir Charles Lyell, published ***Principles of Geology***. His book shook prevailing views on how Earth had been formed. Lyell argued that Earth's surface changed slowly through natural processes. He asserted that it had taken millions of years for layers of sedimentary rock to form. Lyell therefore concluded that Earth was much older than people had previously supposed.

Charles Lyell

DETERMINING THE AGE OF ROCKS

Scientists now use a variety of methods to determine the age of rocks.

LAW OF SUPERPOSITION

If they are studying undisturbed layers of sedimentary rock, scientists assume that the oldest layers of rock are on the bottom and that the youngest layers of rock appear at the top. This is known as the **law of superposition**. Scientists are able to use this knowledge to estimate the age of rocks relative to each other. Scientists also are often able to use evidence from a rock layer to determine the type of environment that existed when the layer was first formed. For example, limestone is formed from the shells of sea creatures. Wherever limestone is found, scientists assume there was once a marine environment.

GRANITE INTRUSION

If there is a break in sedimentary rock, with cross-cutting igneous rock, scientists assume there was a fault in the rock where magma came through — perhaps because of tectonic plate activity. In this case, the igneous rock, known as a **granite intrusion**, is younger than the surrounding sedimentary rock.

LAYERS OF SEDIMENTARY ROCK

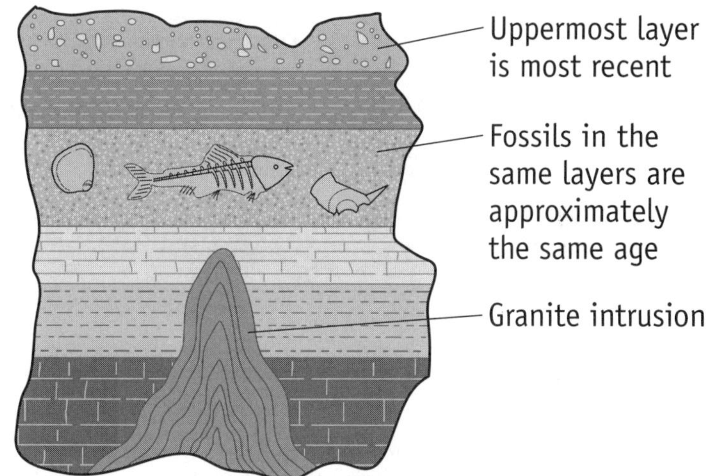

FOSSIL IMPRESSIONS

Under special conditions, dead plants and animals create impressions in sedimentary rocks known as **fossils**. Either a different type of sedimentary material fills an impression left by an organism or minerals gradually seep in and take the place of shells and bones. Based on fossils, scientists can often estimate the age of a layer of sedimentary rock. If they recognize the organism that created the fossil and know when it lived, they can use this knowledge to estimate the age of the rock.

Science Photo Library

Fossil of a fish embedded in a rock

RADIOMETRIC DATING

A more recent and accurate way of dating rocks is **radiometric dating**. This method tells scientists the **absolute age** of different kinds of rocks. Radiometric dating is based on the fact that some elements in rocks are ***radioactive***: these elements are unstable and slowly break down into other elements as they emit radiation. For example, radioactive uranium slowly decays into lead over many millions of years. The speed at which these elements decay is slow but constant and predictable. Scientists are therefore able to estimate the age of a rock by comparing the amount of the radioactive element it contains (*uranium*) with the amount of the substance that this radioactive element becomes after it decays (*lead*). The proportion of these substances in the rock tells them the rock's age.

Geologic time is measured in millions or even in billions of years, divided into **eons**, **eras**, and **periods**. Radiometric dating reveals that Earth is far older than once thought. Scientists now believe that Earth was formed about 4.6 billion years ago. Using radiometric dating, scientists can also tell the age of fossils more accurately. Based on these methods, they now believe the first living organisms appeared on Earth about 4 billion years ago.

TIMELINE OF LIFE ON EARTH

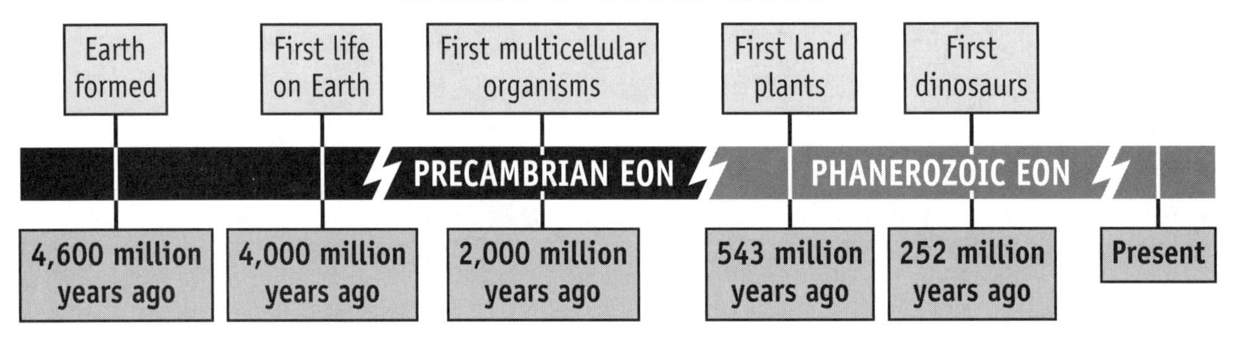

APPLYING WHAT YOU HAVE LEARNED

✦ Describe two ways geologists are able to estimate the age of rocks and fossils.

LIVING ORGANISMS CHANGED EARTH'S ATMOSPHERE

Life on Earth has greatly influenced the planet's systems and processes. An important example of the impact of life on Earth can be seen in the atmosphere. Early in Earth's history, gases from its interior were brought to the surface by volcanoes, creating Earth's primitive atmosphere. These gases included hydrogen, water, ammonia, methane, carbon dioxide, carbon monoxide and nitrogen. However, there was no oxygen in Earth's early atmosphere.

Bacteria and primitive plant life began creating oxygen and absorbing carbon dioxide when they converted sunlight into sugar molecules. Over almost 2 billion years, the amount of oxygen in the atmosphere gradually increased until it reached current levels. Today, Earth's atmosphere is mainly nitrogen (78%) and oxygen (21%).

HUMAN IMPACT ON THE ENVIRONMENT

The appearance of human beings on Earth has also greatly transformed our planet. Humans have cut down forests, irrigated dry plains, created farms and ranches, and built cities. Before the Industrial Revolution, the effects of human activity were small enough to be safely absorbed by Earth's natural processes. However, in the past 200 years, the human population has increased tremendously. Many new technologies are dependent on burning "fossil fuels" (*energy-rich substances that have formed from the decay of long-buried plants and microorganisms*). These technologies threaten Earth's air, water and soil with pollution. Humans are burning up the world's coal and oil, and may one day exhaust these resources.

The burning of fossil fuels is adding carbon dioxide to the atmosphere, heating up Earth through the **"greenhouse" effect:** scientists believe this is responsible for **"global warming."** The carbon dioxide acts as a blanket around Earth. As temperatures rise, the global ice caps are beginning to melt, swelling ocean waters. Other areas are experiencing droughts. Human activity has also stripped Earth of some of its ozone, creating new dangers to all living organisms from solar radiation.

Science Photo Library

Human activities can even alter natural cycles. Tropical rainforests, for example, produce oxygen that enters the atmosphere. To boost farming and lumbering, some countries are cutting down their rainforests. This deforestation may reduce the amount of oxygen in the atmosphere. Both deforestation and monocultural farming (*growing a single crop, often to sell*) also leads to **erosion** and **soil depletion.**

Scientists are trying to reduce the threat of depleting natural resources and destroying the environment by developing alternative energy sources: solar cells can harness the sun's energy to make electricity; windmills, hydroelectric plants and nuclear reactors can produce energy without burning fossil fuels. Scientists and environmentalists are also encouraging the conservation of energy, water, and other precious resources.

APPLYING WHAT YOU HAVE LEARNED

✦ Describe two ways that human activity has had an impact on the environment.

WHAT YOU SHOULD KNOW

★ **Relative Age.** In estimating the relative age of undisturbed layers of sedimentary rock, the oldest layers of the rock are on the bottom and the more recent layers are on the top.

★ **Granite Intrusion.** This occurs when magma has been forced up from below and entered into sedimentary or metamorphic rock.

★ **Radiometric Dating.** To determine the absolute age of a rock, scientists compare the amount of radioactive elements the rock contains with the amount of the substances they become after radioactive decay.

★ **Changes to Earth.** From these methods, scientists estimate the first life appeared on Earth about 4 billion years ago. Early life forms changed Earth's atmosphere over billions of years by adding oxygen.

★ **Dangers to the Environment.** Human activities over the past 200 years now threaten many of Earth's natural systems and processes.

CHAPTER STUDY CARDS

Estimating the Age of Rocks

★ **Law of Superposition**

- In undisturbed sedimentary rock, the oldest layers are on the bottom, and the youngest layers are found at the top.
- Tells the **relative age** of rocks in different layers in the same formation.

★ **Radiometric Dating**

- Tells the **absolute age** of rocks by measuring the decay of radioactive elements.

The Impact of Life on Earth

★ Living organisms first appeared on Earth about 4 billion years ago.

★ Living organisms added oxygen to Earth's atmosphere.

★ Human activity has transformed Earth and threatens many natural processes.

- **Greenhouse Effect.** Burning fossil fuels adds carbon dioxide to the atmosphere, heating up Earth.
- **Global Warming.** Risk of ocean's rising.

CHECKING YOUR UNDERSTANDING

1 A formation of layered rock is directly on top of a formation of non-layered rock. Which of the following would most likely *not* be found in both formations?

A. A granite intrusion that penetrated both rock formations at the same time.

B. Fossils that formed in both formations at the same time.

C. A fault that divided both formations at the same time.

D. Folds from pressures that affected both formations at different times.

- Examine the Question
- Recall What You Know
- Apply What You Know

ES: C 10–3

This question requires that you understand the difference between layered and non-layered rock. Layered rock is **sedimentary**. The non-layered rock could be either igneous or metamorphic. A granite intrusion could go through igneous and metamorphic rock at the same time as sedimentary rock, so Choice A is wrong. The same is true for a fault or folding — Choices **C** and **D**. However, a fossil created by a plant or animal in sedimentary rock cannot also form in either igneous or metamorphic rock. The correct answer is therefore B.

Now try answering other questions about the impact of living organisms on the development of Earth's systems.

2 Why are radioactive substances useful for measuring geologic time?

ES: C 10–3

A. The ratio of decayed products to radioactive substances remains constant.

B. The half-lives of most radioactive substances are short.

C. Samples of radioactive substances are easy to collect from rocks.

D. Radioactive substances undergo decay at a predictable rate.

3 World fossil energy demand is increasing faster today than at any other time in history because of growing economic development by newly emerging industrial superpowers like China and India. Alternative energy sources are essential to avoid serious future energy shortages. Identify one possible alternative energy source and explain how its use might help protect Earth's resources (*2 Points*).

ES: D 10–5

4 Geologic time is a form of absolute time. It relates to when something exactly happened. Relative time relates to the order in which particular events occurred. The diagram below shows a cross-section of five rock layers.

You have been asked to determine the age of rock layer 5. Identify one method that could be used to determine the relative age and another method for determining the absolute age of the rock layer. Describe how each method helps determine the age of the rock layer. (*4 points*)

ES: D 10–5

CHECKLIST OF BENCHMARKS IN THE EARTH AND SPACE SCIENCES UNIT

Directions. Now that you have completed this unit, place a check mark (✔) next to those benchmarks you understand. If you have trouble recalling information connected with one of the benchmarks, review the chapter indicated in the brackets for the items you do not recall.

❑ You should be able to explain how evidence from stars and other celestial objects provide information about the processes that cause changes in the composition and scale of the physical universe. **[Chapter 9]**

❑ You should be able to explain that many processes occur in patterns within Earth's systems. **[Chapter 9]**

❑ You should be able to explain the 4.5 billion-year-history of Earth and the 4 billion-year-history of life on Earth based on observable scientific evidence in the geologic record. **[Chapter 10]**

❑ You should be able to describe the finite nature of Earth's resources and those human activities that can conserve or deplete Earth's resources. **[Chapter 10]**

❑ You should be able to explain the processes that move and shape Earth's surface. **[Chapter 8]**

❑ You should be able to summarize the historical development of scientific theories and ideas, and describe emerging issues in the study of Earth and space sciences. **[Chapters 8 and 9]**

UNIT 4

PHYSICAL SCIENCES

In this unit, you will review what you need to know about the Physical Sciences. You will learn about the nature of matter — including its structure and the properties of different forms of matter. You will also learn about energy, the power that moves matter throughout the universe.

Electricity is one form of energy: light bulbs transform it into light; motors convert it into mechanical energy.

★ **Chapter 11: The Structure of Matter**
In this chapter, you will learn how each atom is made up of the same parts — protons, neutrons and electrons. You will also learn about isotopes and radioactive decay, and how atoms combine with ionic and covalent bonds.

★ **Chapter 12: The Properties of Matter**
This chapter deals with substances and mixtures. Substances have a fixed chemical composition and include compounds and elements, while mixtures are substances that are not chemically combined. You will also learn how to read the Periodic Table of Elements and learn about the properties of different groups of elements, such as metals and nonmetals.

★ **Chapter 13: Force and Motion**
This chapter looks at motion. Sir Isaac Newton realized that objects change their movements because of other forces. From this realization, he developed three laws of motion.

★ **Chapter 14: The Nature of Energy**
Energy comes in many forms. In this chapter, you will learn what energy is, how most energy is either kinetic or potential, how energy is transferred, and how it is transformed from one form to another. You will also learn the properties of energy that travels in waves, such as sound or light waves.

CHAPTER 11

THE STRUCTURE OF MATTER

In this chapter, you will learn about atoms, and how they combine to form ions and molecules.

ATOMS AND SUBATOMIC PARTICLES

MAJOR IDEAS

A. All matter is made up of minute particles known as atoms. Atoms are composed of protons, neutrons, and electrons.

B. Different isotopes of the same element have different numbers of neutrons.

C. Atoms react with other atoms to obtain a complete outer energy level. Atoms ***transfer*** electrons to form ionic bonds. Atoms share electrons to form covalent bonds, which hold together molecules.

D. Both ***before*** and ***after*** a chemical reaction, the number of atoms and total mass is the same.

Anything that has **mass** and occupies space is known as **matter**. Air, glass and water are matter. Light and electricity, however, are not matter. All matter is made up of minute particles called atoms. An **atom** is the smallest part of an element that has the physical and chemical properties of that element. An **element** is a substance with only one kind of atom. For example, all aluminum is made up of aluminum atoms.

THE PARTS OF THE ATOM

Scientists once thought that atoms were indivisible. However, in the 20th century, scientists discovered that atoms are made up of even smaller **subatomic** particles. The three most important subatomic particles found in all atoms (*except hydrogen atoms, which lack neutrons*) are:

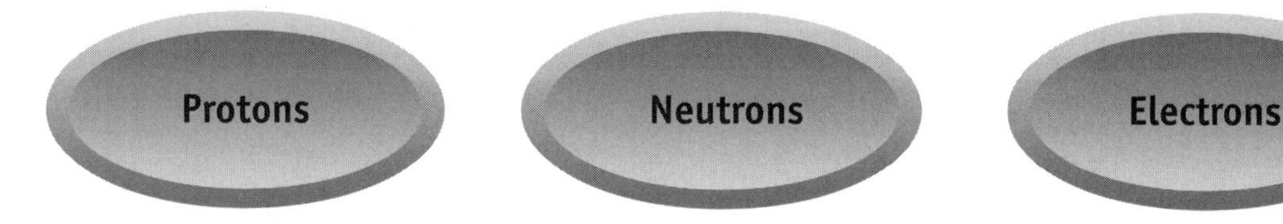

★ The **proton** is a particle with a positive charge and a mass of 1.673 X 10^{-24} grams, or **1 atomic mass unit**. Protons are located in the **nucleus** of the atom. The nucleus is at the dense center of the atom.

★ The **neutron** is a neutral particle with approximately the same mass as the proton (*1 atomic mass unit*). Neutrons are also located in the nucleus.

★ **Electrons** are small, negatively charged particles that move around the nucleus at very high speeds. Compared to protons and neutrons, they have no significant mass. They are attracted to the positively charged protons in the nucleus. No one can actually measure the speed, direction, and location of an electron. However, scientists believe electrons occupy different **energy levels** or **shells**, based on the way atoms behave. The first energy level can hold up to two electrons. The second energy level can hold up to eight electrons. You will learn more about these energy levels, or shells, later in this chapter.

Atoms are so small they cannot be seen, even by a powerful microscope. However, scientists often try to picture atoms with diagrams that explain how they behave: protons and neutrons are shown together in the nucleus. Electrons are shown circling the nucleus in orbits, or more accurately, as **electron clouds**. The following illustrations help us understand how atoms act. In reality, the electrons are actually much smaller and farther from the nucleus than shown here.

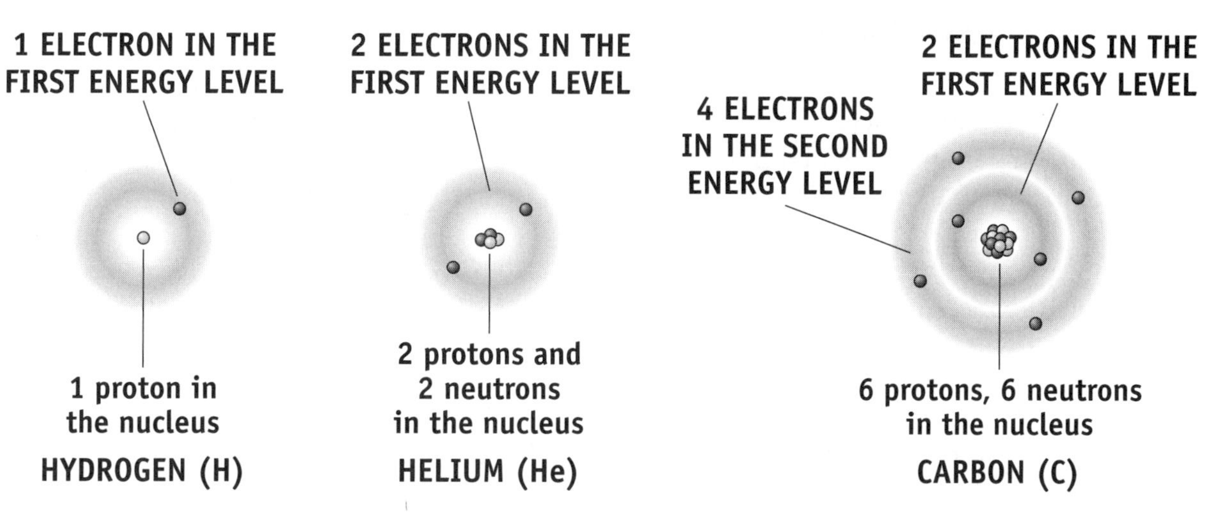

★ **Electrically Neutral.** Atoms have the same number of positively charged protons and negatively charged electrons. They therefore have no electrical charge and are considered neutral.

★ **Atomic Symbols.** To identify each element, scientists use a symbol of one or two letters, based on its name. The first letter is always capitalized, but not the second. For example, the symbol for hydrogen is **H**. The symbol for helium is **He.**

★ **Atomic Numbers.** Every element has its own unique atomic number. The number of protons an atom of the element has determines its atomic number. In the example on the previous page, helium has an atomic number of 2, and carbon has an atomic number of 6. The atomic number also reveals the number of electrons, since these are the same as the number of protons.

★ **Atomic Mass.** To determine the mass of an atom, add the number of its protons and neutrons. The mass of a typical hydrogen atom is 1, of a typical helium atom is 4, and of a typical carbon atom is 12. If you know the atomic mass and the atomic number of an atom, you can determine the number of its protons, neutrons, and electrons.

ATOMIC NUMBER = NUMBER OF PROTONS = NUMBER OF ELECTRONS
ATOMIC MASS – ATOMIC NUMBER = NUMBER OF NEUTONS

Scientists often write the atomic mass of the element on the upper left of the symbol, and the atomic number on the lower left.

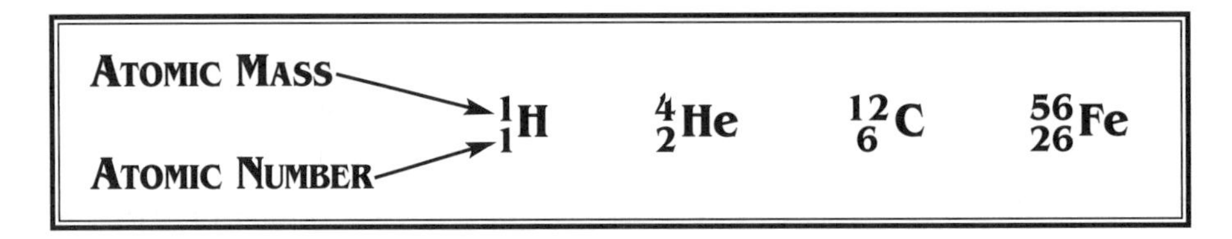

ISOTOPES AND RADIOACTIVE DECAY

Atoms of the same element always contain the same number of protons, but the number of their neutrons may actually vary. This gives some atoms of the same element different masses. Atoms of the same element with different masses (*different numbers of neutrons*) are called **isotopes**. For example, there are two different isotopes for carbon: $^{12}_{6}\mathbf{C}$ and $^{14}_{6}\mathbf{C}$

Both carbon isotopes have six protons and six electrons, but ^{14}C (or C-14) has eight neutrons, while the more common ^{12}C (or C-12) has only six neutrons. Differences among isotopes explain why the atomic masses of most elements on the **Periodic Table** contain a decimal. The atomic mass for each element on the Periodic Table is based on all known isotopes of that element. Each isotope is given weight based on how common it is in nature. Thus, the atomic mass for carbon (C) is 12.0107. You will learn more about the Periodic Table in the next chapter.

Radioactivity. The atoms of some elements have nuclei that are unstable. The force holding the nucleus together isn't strong enough to overcome the electrical force pushing the protons apart. The nuclei of these atoms break apart in a process of spontaneous decay known as **radioactivity**. Sometimes different isotopes of the same element are either stable or radioactive. Radioactive substances undergo nuclear decay by emitting both particles (*such as protons*) and high-energy, wavelike radiation. When a radioactive substance decays, a new element is often formed. This processes is known as **transmutation**. The decay of radioactive substances takes place at fixed and predictable rates. The time in which it takes half of a radioactive substance to decay is known as its **half-life**.

APPLYING WHAT YOU HAVE LEARNED

✦ An atom of nickel has an atomic number of 28 and an atomic mass of 59. How many protons, neutrons, and electrons does it have?

✦ What are the differences between these two atoms? $^{35}_{17}Cl$ and $^{37}_{17}Cl$

WHAT YOU SHOULD KNOW

A. All matter is made up of minute particles known as atoms. Atoms are made up of even smaller **subatomic** particles known as protons, neutrons, and electrons.

B. **Protons** are located in the dense nucleus of the atom. The proton is a particle with a positive charge. **Neutrons**, also located in the nucleus, are neutral. **Electrons** are small, negatively charged particles that move around the nucleus at high speeds.

C. Atoms have the same number of positively charged protons and negatively charged electrons. To identify each element, scientists use a symbol of one or two letters.

D. Atoms of the same element with different masses (*different numbers of neutrons*) are called **isotopes**.

E. The atoms of some elements have nuclei that are unstable. The nuclei of these atoms break apart in a process of spontaneous decay known as **radioactivity**. The time in which it takes half of a radioactive substance to decay is known as its **half-life**.

IONS AND MOLECULES

Atoms react with each other to form ions and molecules. Electrons play a key role in this process. This often results in new substances with their own physical and chemical properties.

IONIC AND COVALENT BONDING

Electrical charges may be positive (+) or negative (-). Opposite electrical charges attract and similar charges repel each other. The attraction between the positively charged nucleus and negatively charged electrons holds an atom together. The energy of the electrons moving around the nucleus prevents the electrons from collapsing into the nucleus. The location of the electron around the nucleus depends on the amount of energy that the electron has and on the attractive force that the nucleus has for that electron.

ENERGY LEVELS

Electrons can only exist in particular **energy levels** around the nucleus. Each level can hold only a specific number of electrons. As the distance from the nucleus increases, the number of the electrons that each energy level can hold increases and the attraction of the electrons to the nucleus decreases.

You can think of these energy levels as similar to a parking lot, in which all the rows in the lot do not hold the same number of cars. The rows closest to the exit are the ones that fill up first. Similarly, energy levels generally fill from the nucleus outward. The first energy level, closest to the nucleus, can hold up to two electrons. Any additional electrons must move out to the second energy level. This energy level can hold up to 8 electrons. The third energy level can hold up to 18 electrons, and the fourth energy level can hold up to 32 electrons.

Level	Number of Electrons
1	2
2	8
3	18
4	32

Valence Electrons. The electrons in the outermost energy level of an atom are known as its **valence electrons**. An atom is always most stable when it has filled its outermost energy level. Atoms often act either to lose extra electrons or to attract additional electrons to fill this outermost energy level. They can do this by either transferring or sharing electrons with other atoms. This tendency helps determine many of an atom's chemical properties.

IONIC BONDS

One way that atoms can achieve a more stable arrangement of electrons in their outermost energy level is by transferring electrons. When an atom gains or loses one or more electrons, it is known as an **ion**. Unlike an atom, an ion has an electrical charge. When atoms lose electrons, they form positively charged ions. When they gain electrons, they form negatively charged ions.

When the transfer of electrons occurs between two or more atoms, the negatively charged ions become attracted to the positively charged ions. An **ionic bond** forms between the positively and negatively charged ions, locking them into place. Ionic bonds occur between a metal and nonmetal.

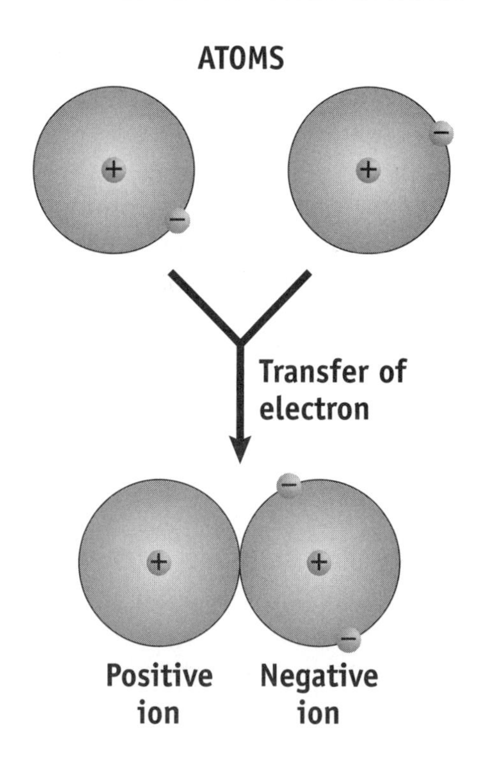

For example, an atom of magnesium will ***lose two electrons*** to become a Mg^{2+} ion. An atom of chlorine has seven electrons in its outer energy level. One more electron would complete this outer level. It will ***gain one electron*** to form a Cl^{-} ion. Because chlorine lacks **one** electron in its outer level and magnesium has **two** extra electrons in its outer level, two chlorine atoms are needed for each magnesium atom. Each chlorine atom takes one electron from the magnesium. The oppositely charged ions then attract each other to form ionic bonds. The positively charged magnesium ion becomes locked into place by its attraction to the two negatively charged chloride (*chlorine*) ions. One atom of magnesium forms ionic bonds with two atoms of chlorine, forming **$MgCl_2$** (*magnesium chloride*). These atoms are always found in the same ratio.

COVALENT BONDING

In other cases, atoms complete their outer energy levels by *sharing* electrons. The same pair of electrons will actually move around the nuclei of both atoms. This is known as a **covalent bond**. Covalent bonds occur between nonmetals. This type of bond differs from ionic bonding.

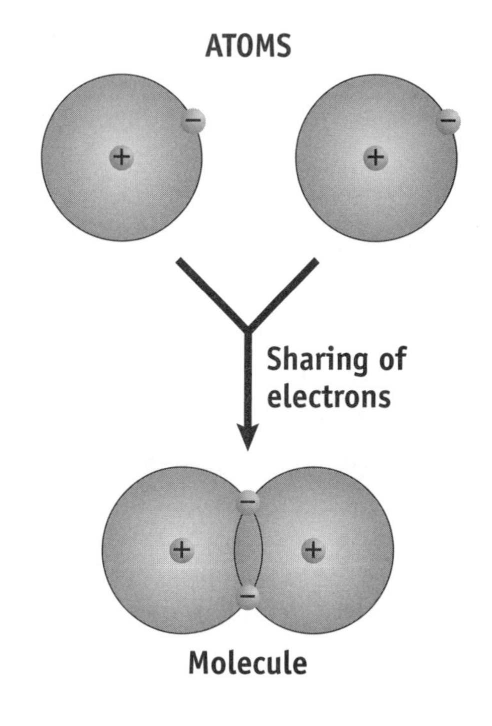

When electrons are shared by several atoms, a **molecule** is formed. Covalent bonds hold the molecule together. For example, in a molecule of oxygen, two atoms of oxygen share two pairs of electrons to fill their outermost energy levels. The result is an oxygen molecule (**O_2**). The molecule is held together by a covalent bond because the electrons are shared.

The number of electron pairs that need to be shared depends upon the number of electrons needed to complete each atom's outer energy level. Covalent bonds may be formed in molecules containing more than two atoms. In a water molecule (**H_2O**), one oxygen atom shares a pair of electrons with two hydrogen atoms. Each hydrogen atom ends up with two electrons in its outermost energy level, while the oxygen atom ends up with eight valance electrons in its outer level.

CHEMICAL REACTIONS

In a chemical reaction, two or more substances combine to form new substances with different properties. The number and types of atoms and the total mass remain the same ***before*** and ***after*** the reaction. For example, when a magnesium atom forms ionic bonds with two chlorine atoms, the reaction is: $\mathbf{Mg + 2Cl \rightarrow Mg\ Cl_2}$. There are two chlorine atoms and one magnesium atom both before and after the reaction.

When two hydrogen molecules, ($\mathbf{2H_2}$) form covalent bonds with one oxygen molecule, ($\mathbf{O_2}$), the chemical reaction is:

$$2H_2 + O_2 \rightarrow 2H_2\,O$$

If you examine this equation, the number of atoms at the beginning of the reaction is again the same as the number of atoms at its completion. On both sides of the equation, there are the same six atoms: four hydrogen atoms and two oxygen atoms. However, they are combined differently. On the left of the equation, they form three gas molecules. On the right side, the atoms have combined into two water molecules. The gases on the right side of the equation are both **elements**: the molecules are made up of the same kinds of atoms. Water is a **compound**: it has different kinds of atoms combined together. Although water has both hydrogen and oxygen atoms, it is a completely different substance with its own chemical and physical properties.

UNDERSTANDING BALANCED CHEMICAL EQUATIONS

$$2H_2 + O_2 \rightarrow 2\,H_20$$

When scientists write chemical formulas and equations, they follow certain rules:

- ★ The ***reactants*** are usually written on the left side of the reaction. The arrow indicates the direction of the reaction. Here, oxygen and hydrogen combine to form water. The ***products*** are written on the right side after the arrow.
- ★ To indicate the number of **atoms** in each molecule, they put a number below the line after the atomic symbol. In $\mathbf{H_2}$, the small number $_2$ indicates that the hydrogen molecule has two atoms. If there is no number after the atomic symbol, then the number is assumed to be one. In the water molecule H_20, there is only one oxygen atom.
- ★ To indicate the number of **molecules** involved in the reaction, scientists write the number in front of the molecule. In this example, "$\mathbf{2H_2}$" indicates that there are two hydrogen molecules. When there is no number in front of a molecule, the number of molecules is one. In ($\mathbf{2H_2 + O_2 \rightarrow 2\ H_2O}$), there is only one oxygen molecule on the left side of the equation.
- ★ The equation is **balanced** if there are the same number of atoms on each side of the equation. Multiply the number of molecules by the number of atoms of each type to check. Here there are four hydrogen and two oxygen atoms on each side, so the equation is balanced.

THE CONSERVATION OF MASS AND ENERGY

The total mass of the reactants before a chemical reaction and the mass of the products after the reaction must always be the same. This illustrates the principle known as the **conservation of mass**. Mass cannot be created or destroyed in a chemical reaction or by other ordinary means.

A chemical reaction will also either require or produce energy. This energy is stored or released by the ionic or covalent bonds. If energy is released, it will usually be spread to the surroundings in the form of heat. This energy does not go away, it just spreads to neighboring matter. If energy is required, it must be supplied by some outside source. Energy, like matter, also cannot be created or destroyed, although it can be stored in chemical bonds. This principle is known as the **conservation of energy**.

APPLYING WHAT YOU HAVE LEARNED

- ✦ How does an ionic bond differ from a covalent bond?
- ✦ Explain the principle of conservation of mass.
- ✦ Methane (CH_4) is combined with oxygen (O_2) to produce carbon dioxide (CO_2) and water (H_2O). Write a balanced chemical equation representing this reaction.

WHAT YOU SHOULD KNOW

- ★ Opposite electrical charges attract and similar charges repel each other. The attraction between the positively charged nucleus and negatively charged electrons holds an atom together.
- ★ Electrons can only exist in particular **energy levels** around the nucleus. Each energy level can hold only a specific number of electrons.
- ★ The electrons in the outermost energy level of an atom are known as its **valence electrons**. An atom is always most stable when it has filled its outermost energy level.
- ★ Some atoms complete their outermost energy levels by transferring electrons and becoming ions. An **ionic bond** forms between the positively and negatively charged ions. Some atoms complete their outermost energy level by *sharing* electrons. The same pair of electrons move around the nuclei of both atoms, forming a **covalent bond**.
- ★ In a chemical reaction, two or more substances combine to change their chemical properties. The number and type of atoms and the total mass must be the same b***efore*** and ***after*** the reaction. This demonstrates the conservation of mass.

CHAPTER STUDY CARDS

Atoms and Subatomic Particles

Atom. Smallest unit of matter unique to a particular element. Atoms contain:

- ★ **Protons.** Positively charged; all atoms have protons; have atomic mass; located in the nucleus of the atom.
- ★ **Neutrons.** Neutral in charge; the same mass as the proton; also located in the nucleus of the atom.
- ★ **Electrons.** Negatively charged; attracted to nucleus' positive charge; move around the nucleus at high speeds in different energy levels; same number as protons.

Atomic Notation

- ★ **Atomic Symbol.** Scientists use a symbol of one or two letters to identify elements.
- ★ **Atomic Number** = number of protons.
- ★ **Atomic Mass** = the number of protons and neutrons.

Radiation

- ★ **Isotopes.** Atoms of the same element with different numbers of neutrons.
- ★ **Radioactive Decay.** Some elements or isotopes of elements have unstable nuclei, which break apart, causing radioactivity.

Ions and Molecules

- ★ **Energy Levels.** Electrons can only exist in particular **energy levels** around the nucleus. Each level can hold only a specific number of electrons. An atom is most stable when it fills its outermost energy level
- ★ **Ionic Bonds.** When a transfer of electrons occurs between two or more atoms, negatively charged ions are attracted to positively charged ions.
- ★ **Covalent Bonds.** Some atoms (nonmetals) complete their outer levels by sharing electrons. The same pair of electrons move around the nuclei of both atoms in covalent bonding.

Chemical Reactions

- ★ **Chemical Reactions.** These are combinations of substances that result in one or more new substances with new physical and chemical properties.
- ★ **Balancing Equations.** There should be the same number of atoms on each side of the equation. Multiply the atoms in each molecule by the number of molecules to see that they balance: $2H_2 + O_2 \rightarrow 2\ H_2O$
- ★ **Conservation of Mass.** Matter cannot be created or destroyed in a chemical reaction.

CHECKING YOUR UNDERSTANDING

1 Which statement correctly describes protons and neutrons?

PS:A 9–1

A. They have the same mass and the same electrical charge.

B. They have the same mass but different electrical charges.

C. They have different masses but the same electrical charge.

D. They have different masses and different electrical charges.

This question tests your understanding of subatomic particles. Both protons and neutrons are located in the nucleus. Each proton has an atomic mass of one and a *positive* electrical charge. Each neutron has an atomic mass unit of one and *no electrical charge*. Only choice B is correct in stating that they both have the same masses but different electrical charges.

Now try answering some additional questions about the structure and properties of atoms and subatomic particles.

Use the table that follows to answer questions 2 through 4:

DATA TABLE

Substance	Number of Protons	Number of Electrons
Sulphur	16	18
Chromium	24	24
Cobalt	27	24
Americium	95	92

2 **Which of these substances is electrically neutral?**

A. sulphur
B. chromium
C. cobalt
D. americium

- ♦ Examine the Question
- ♦ Recall What You Know
- ♦ Apply What You Know

PS: A 9–2

3 **Which statement is true about the electrical charges assigned to the electrons and protons of the elements in the table?**

A. The electrons and protons are both positive.
B. The electrons are positive and the protons are negative.
C. The electrons are negative and the protons are positive.
D. The electrons and protons are both negative.

PS: A 9–2

4 **Which of the substances on the table is a negatively charged ion?**

A. sulphur
B. chromium
C. cobalt
D. americium

PS: B 9–5

5 Which spontaneously emits radiation and particles from the nuclei of its atoms?

A. inert gases
B. radioactive isotopes
C. chemical reactions
D. covalent bonds

6 Which force holds atoms together?

A. attractive electrical forces between protons and electrons
B. magnetic forces of attraction
C. frictional forces between protons and electrons
D. gravitational force

- Examine the Question
- Recall What You Know
- Apply What You Know

PS: A 9–6

7 The composition of the nucleus of carbon varies from one isotope to another. The symbol for one isotope of carbon is $^{14}_{6}C$. How many neutrons are contained in this isotope?

A. 6
B. 8
C. 14
D. 20

PS: A 9–1

Use the chart below to answer the following question.

	The number of electrons at each energy level		
Element	**1st Level**	**2nd Level**	**3rd Level**
Magnesium(Mg)	2	8	2
Oxygen (O)	2	6	

8 When Mg forms an ionic bond with O, how many electrons are transferred from the Mg atom to the O atom?

A. 2
B. 6
C. 12
D. none

PS: B 9–5

---

9 Which of the following statements describes electrons?

A. They are positive subatomic particles located in the nucleus.
B. They are positive subatomic particles that move around the nucleus.
C. They are negative subatomic particles located in the nucleus.
D. They are negative subatomic particles that move around the nucleus.

PS: A 9–2

10 Which of the following represents a pair of isotopes?

A. ^{1}H and ^{3}H
B. $^{40}K^{2-}$ and $^{40}Ca^{-}$
C. $^{16}O^{2-}$ and $^{19}F^{1-}$
D. $^{16}O^{2-}$ and $^{32}S^{2-}$

◆ Examine the Question
◆ Recall What You Know
◆ Apply What You Know

PS: A 9–1

11 An atom contains 12 neutrons and 11 electrons. What is the number of protons in this atom?

A. 1
B. 11
C. 12
D. 23

PS: A 9–1

12 Different isotopes of the same element have different

A. atomic numbers.
B. numbers of neutrons.
C. numbers of protons.
D. numbers of electrons.

PS: A 9–1

13 If the atom of an element has an atomic number of 15 and its atomic mass is 31, how many neutrons does it have?

A. 15
B. 16
C. 31
D. none

PS: A 9–1

14 Covalent bonds are formed when electrons are

A. transferred from one atom to another.
B. mobile within a metal.
C. captured by the nucleus.
D. shared between two or more atoms.

PS: B 9–7

15 A scientist combines hydrogen chloride (HCl) and sodium hydrochloride (NaOH) to produce table salt (NaCl) and water (H_2O). Which equation best represents what took place?

A. $2NaOH \rightarrow HCl \rightarrow 2H_2O + NaCl$
B. $NaOH + HCl \rightarrow H_2O + 2NaCl$
C. $NaOH + HCl \rightarrow H_2O + NaCl$
D. $2NaOH + 2HCl \rightarrow H_2O + NaCl$

PS: B 9–7

CHAPTER 12

THE PROPERTIES OF MATTER

In this chapter, you will learn about the properties of substances and mixtures, such as how they look or react. You will also learn about the Periodic Table of the Elements, which classifies elements based on their properties.

MAJOR IDEAS

A. A **substance** is homogeneous with a fixed chemical composition. A **mixture** contains two or more substances mixed together without being chemically combined.

B. The **physical properties** of any substance or mixture include color, odor, density, melting point, boiling point and the ability to conduct electricity. **Chemical properties** refer to the ability of a substance or mixture to react with other matter.

C. The **Periodic Table of the Elements** lists elements by their **atomic number** (*number of protons*). The table is arranged in horizontal rows called **periods** and vertical columns called **groups**.

D. An element's location on the **Periodic Table of the Elements** indicates its electron arrangement and general properties. Metals, metalloids, nonmetals, and noble gases are listed in the Periodic Table.

E. Matter can exist as a **solid**, **liquid**, or **gas**. Matter does not change its chemical composition when it changes its state.

F. The pH of a solution (0-14) tells how acidic or basic it is.

G. Substances and mixtures can conduct electricity if they have freely moving electrons.

SUBSTANCES AND MIXTURES

A **substance** is any kind of matter that is **homogeneous** (*the same throughout, or uniform*) and has a fixed chemical composition.

Elements and compounds are both substances. An **element** is a substance in which all the atoms are the same — for example, mercury (**Hg**) or oxygen (**O_2**). **Compounds** are substances that contain atoms of different types — for example, water (**H_20**), or salt (**NaCl**). Although a compound has different types of atoms, their proportions are fixed because these atoms are chemically combined by covalent or ionic bonds. Compounds always have atoms of each element in the same proportion. For example, water is a compound that always has one oxygen atom for every two hydrogen atoms. The elements in a compound cannot be separated without a chemical reaction.

Lawrence Livermore Laboratories

A student carefully combines substances.

A **mixture** contains two or more substances that are mixed together without being chemically combined. The composition of mixtures can be varied. For example, salt water is a mixture because the amount of salt or water in salt water can easily be changed. There is no fixed proportion between the ingredients in the mixture. The substances also retain many of their original properties — salt water tastes salty. Finally, the substances in a mixture can be separated without a chemical reaction. If a salt water solution is boiled, the water will evaporate and the salt will remain.

Element: Sodium (Na)	Compound: Salt (NaCl)	Mixture: Salt water (NaCl, H_2O)
All the atoms are the same.	Different kinds of atoms are uniformly arranged throughout in fixed amounts; has its own unique properties; cannot be separated without a chemical reaction.	Several substances without being chemically combined; can be separated without a chemical reaction; substances keep some of their properties.

Homogeneous and Heterogeneous Mixtures. There are two types of mixtures: **homogeneous** (*uniform*) and **heterogeneous** (*non-uniform*). **Solutions** are homogeneous. In a solution, small particles of one substance are evenly spread among the particles of the other substance. For example, salt water is a solution. At some point, a solution cannot dissolve any more particles. At this point, the solution is **saturated**.

A solution of salt water is a homogeneous mixture because the amount of each substance (*water and salt*) is uniform throughout the solution. In solid form, salt crystals are held together by the attraction between their **Na^+** and **Cl^-** ions. A water molecule has a negative side (*oxygen*) and a positive side (*hydrogen*). Salt crystals will dissolve in water because the electrical charges of the water molecules pull apart the ionic bonds of the salt crystals. The **Na^+** and **Cl^-** ions are separated, surrounded by water molecules, and equally dispersed throughout the solution.

Some homogeneous mixtures are solids. An **alloy** is a homogeneous mixture made of two metals. The metals in an alloy have been melted, mixed, and allowed to cool. On the other hand, a spoonful of salt and pepper forms a heterogeneous mixture. The salt and pepper are not spread uniformly throughout the mixture.

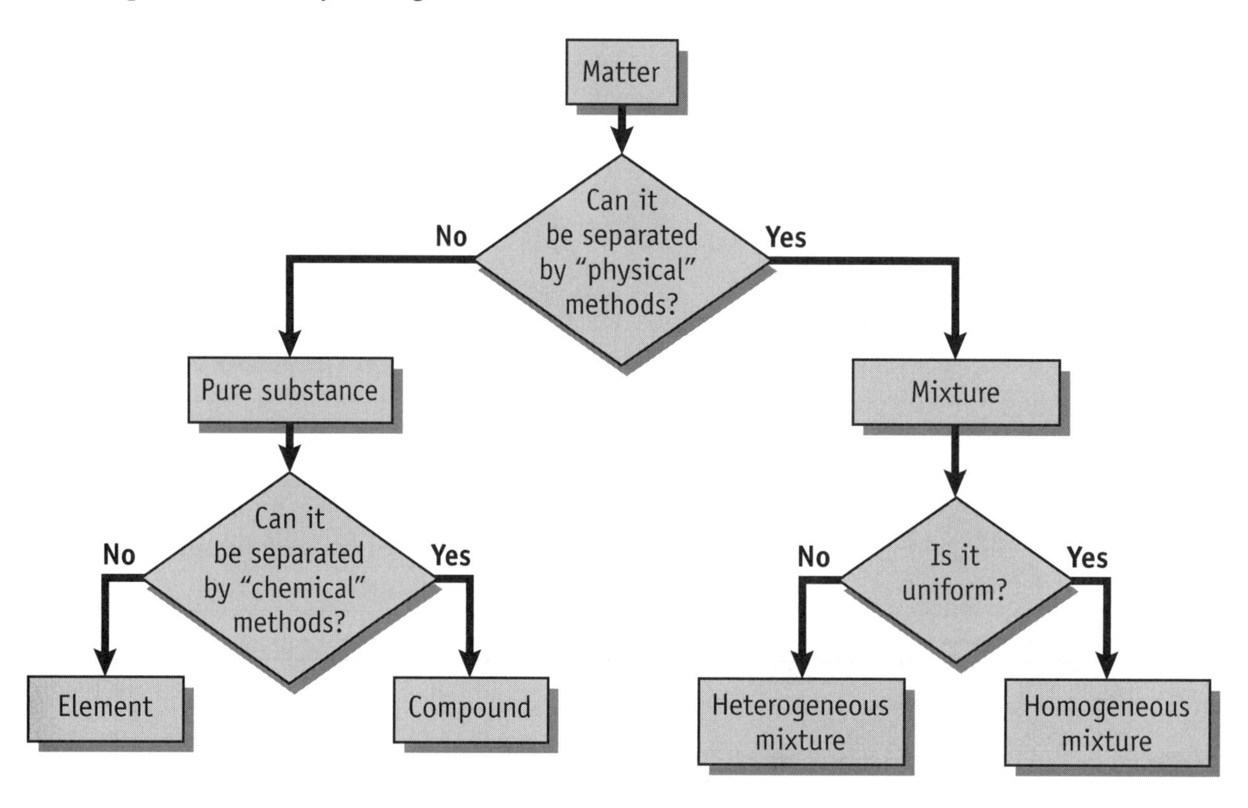

PHYSICAL AND CHEMICAL PROPERTIES

The **physical properties** of matter — either a substance or a mixture — include color, odor, density (mass/volume), hardness, melting and boiling point, and the ability to conduct heat and electricity.

Chemical properties refer to the ability of one form of matter to react with other forms of matter. For example, some substances are flammable — if heated, they will react with oxygen in combustion. The flammability of a substance is a chemical property.

THE PERIODIC TABLE OF THE ELEMENTS

Science Photo Library

Dmitri Mendeleev

In the mid-1800s, **Dmitri Mendeleev**, a Russian chemist, noticed repeating patterns in the chemical properties of the elements known at that time. Mendeleev developed a table based on those patterns, and structured his table by atomic mass. When many blank spaces existed in his table, he theorized these spaces were for yet undiscovered elements. His periodic table led to the discovery of many new elements.

Since Mendeleev, the **Periodic Table of the Elements** has been modified to list elements by their atomic number (*number of protons*). The table is arranged in horizontal rows called **periods**. The period tells you how many energy levels the atom has. Each period begins with an atom containing only one valence electron and ends with an atom with a complete outer level containing eight electrons. The vertical columns of the Periodic Table are called **groups** or **families**. The members of each group have similar chemical and physical properties since they contain the same number of valence electrons.

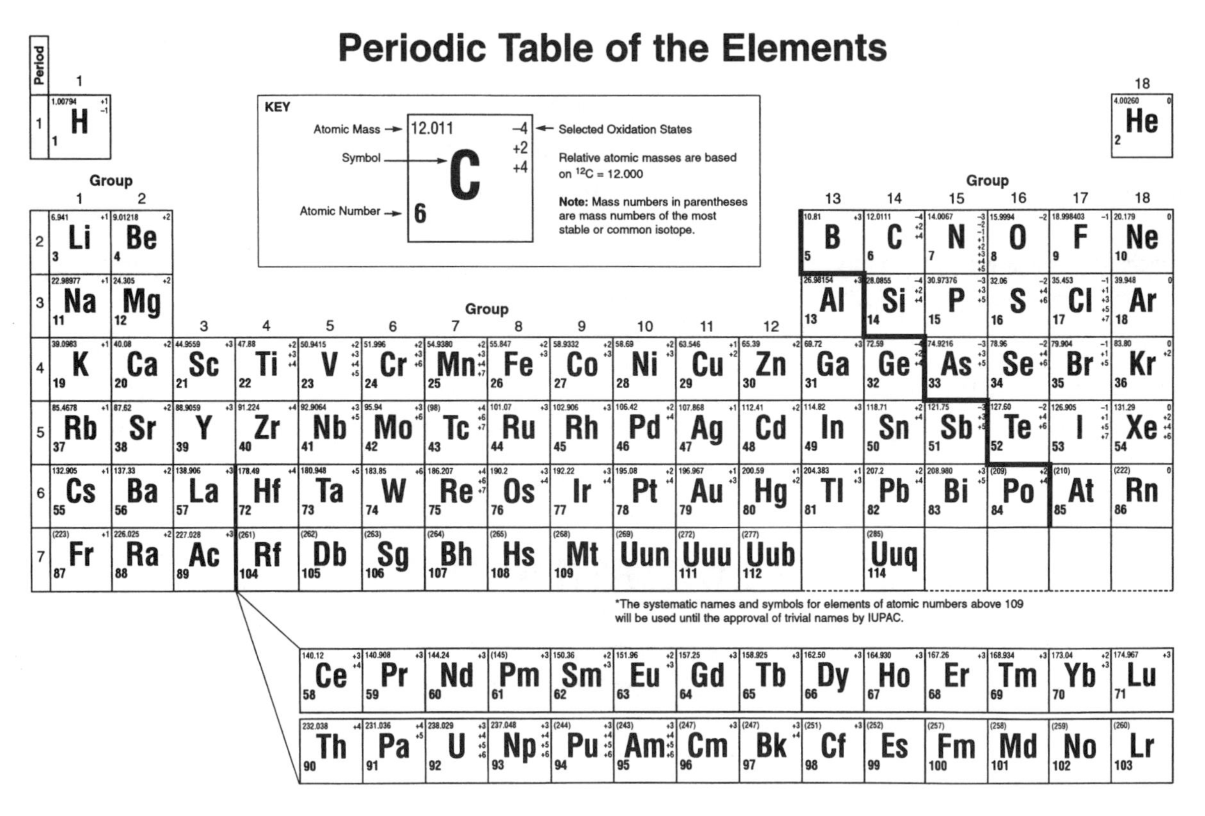

THE ARRANGEMENT OF ELECTRONS

You can tell the arrangement of an element's electrons and predict how it will combine from its location on the table. For example, Group 1 has one electron in its outer energy level. Period 1 has only one energy level. Thus, hydrogen has just one electron. Lithium (**Li**) is in Period 2. It has two energy levels. Because it is in Group 1, we know it has one electron in its outer level. This means a lithium atom has two electrons in its first energy level, and one electron in its second energy level.

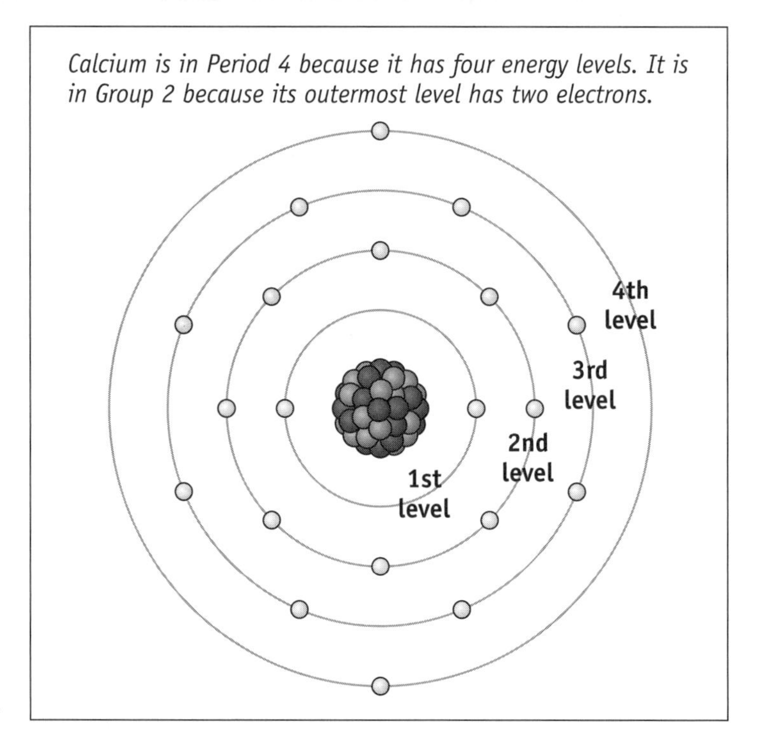

Calcium is in Period 4 because it has four energy levels. It is in Group 2 because its outermost level has two electrons.

Can you tell the electron arrangement for calcium (**Ca**), which is in Group 2 and Period 4? A calcium atom has two electrons in its first energy level, eight electrons in its second energy level, eight electrons in its third energy level, and two electrons in its fourth energy level. This makes 20 electrons in all.

For Groups 13 to 18, Group 13 has three electrons in its outermost energy level; Group 14 has four; and so on. Groups 3-12 are transition metals, which follow slightly different rules.

APPLYING WHAT YOU HAVE LEARNED

✦ Using the Periodic Table of the Elements found on page 113, make a diagram of a nitrogen (N) atom. Show its energy levels with their electrons; protons; and neutrons.

GENERAL PROPERTIES OF ELEMENTS

From the Periodic Table, we can see that the repeating patterns of many physical and chemical properties of elements, including how they combine, are due to their electron configurations.

METALS

Metals, the largest group of elements, are found on the left side of the table. Metals lose electrons and form positively charged ions when they react with other elements.

The atoms of metals are packed tightly together. Their outer energy levels overlap, making it possible for their electrons to move from atom to atom. Metals are solid at room temperature (*except for mercury*), shiny, good conductors of electricity and heat, ductile (*can be drawn into thin wires*), and malleable (*can easily be hammered into thin sheets*).

Alkali Metals. The elements of Group 1, the column to the very left of the table, (*see page 113*) are known as the **alkali metals** (*for example, lithium, sodium, and potassium*). Because they have only one valence electron, which can be easily removed, the alkali metals are highly reactive. They are almost never found in nature as pure elements.

Science Photo Library

Lithium, sodium, and potassium, clockwise from top left

METALLOIDS

The six elements that border the bolded "staircase line" on the table (*boron, silicon, germanium, arsenic, antimony, tellurium, and polonium*) are referred to as **metalloids** or **semiconductors**. They exhibit some of the properties of both metals and nonmetals. The metalloids are economically important because they partially conduct electricity. They are especially valuable in the semiconductor and computer chip industry. Silicon is a metalloid element used in the manufacture of computer chips.

NONMETALS

To the right of the bolded staircase are the nonmetals (*for example, carbon, nitrogen, oxygen, phosphorus, sulfur, and selenium*). Nonmetals are basically the opposite of metals. They *gain* electrons to fill their outermost energy level, and therefore form negative ions where they combine with metals. They can also form covalent bonds with other nonmetals. Nonmetal solids like carbon and sulfur are brittle, non-ductile, non-malleable, and do not conduct electricity. Nonmetals have no metallic luster and do not reflect light. Many nonmetals, such as oxygen and nitrogen, are found in nature as gases. Group 17, the **halogens**, are highly reactive.

NOBLE GASES

Group 18 at the far right of the table, consists of the **noble gases** (*helium, neon, argon, krypton, xenon, and radon*). These elements are generally very unreactive or inert because they have the maximum number of electrons (8) possible in their outer shells. For example, argon (Ar) is often used to fill light bulbs because it does not easily react. In nature, the noble gases often exist as single atoms.

SUMMARY: TYPES OF ELEMENTS

	Metals	Alkali Metals	Metalloids	Nonmetals	Noble Gases
Examples	Sodium (Na), gold (Au), iron (Fe), silver (Ag)	Lithium (Li), sodium (Na), magnesium (Ma)	Boron (B), silicon (Si), arsenic (As)	Carbon (C), sulfur (S), oxygen (O), nitrogen (N)	Helium (He), neon (Ne), argon (Ar), xenon (Xe)
Electrons	Give up electrons when bonding	Form of metal; gives up one electron	May give up or accept electrons	Accepts electrons when bonding	Completed outermost energy shell
Properties	Good conductors of electricity and heat; high melting and boiling points; hard, shiny, and ductile	Rarely found uncombined in nature; highly reactive	Share properties of metals and nonmetals; conduct electricty under some conditions	Poor conductors of heat; do not conduct electricity; solids are brittle, non-ductile; several are gases	Not reactive

APPLYING WHAT YOU HAVE LEARNED

- ✦ Explain the effect that having a completed outermost energy level has on the noble gases.
- ✦ Compare the properties of metals and non-metals.

THE THREE STATES OF MATTER

Matter may exist in either a ***solid***, ***liquid***, or ***gaseous state***. One important characteristic of different forms of matter is their melting and boiling points. These are the points at which matter changes state. Matter does not change its chemical composition when it changes its state.

Solids. Scientists believe that all atoms, ions, and molecules are in constant motion. In a solid, atoms, ions, and molecules are locked into fixed positions, often in a network like a crystal. This gives the substance both fixed volume and a fixed shape. The ions and molecules vibrate in place.

WATER

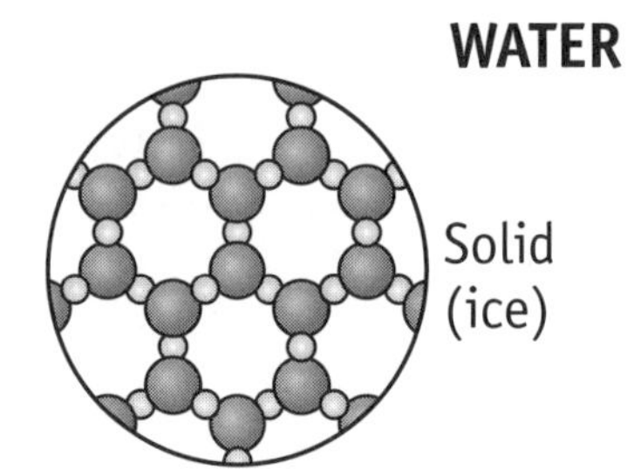

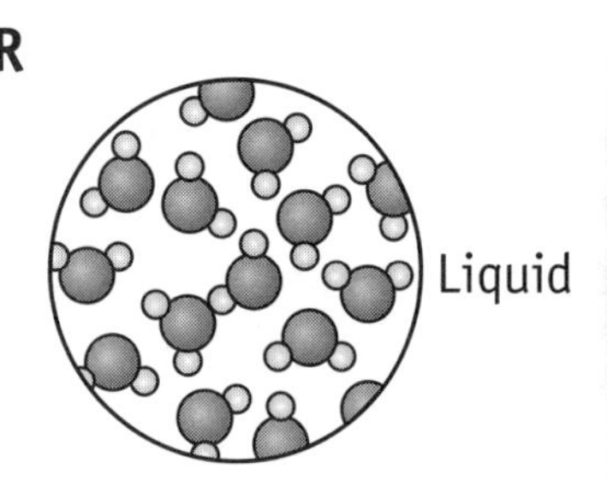

Liquids. When heat energy is transferred to a substance or mixture, its particles begin to vibrate more rapidly. Eventually, the ions or molecules vibrate so strongly they break away from their position and start to move around each other. The solid melts into a liquid. The temperature at which a solid turns into a liquid is known as its **melting point.**

As a liquid, the substance or mixture still has a fixed volume, but not a fixed shape. Its particles are not as tightly packed as in a solid state so its volume is usually larger. Because the particles of a liquid can easily move around each other, the liquid will take the shape of whatever container it is in.

Gas. If yet more heat is applied to the liquid, its ions and molecules will move around even more rapidly. Eventually, they break all connections with the other ions or molecules of the substance or mixture and spread out in all directions as a gas. Gas molecules are spread further apart than in liquids or solids. A gas has no fixed shape or volume. The temperature at which a liquid turns into a gas is known as its **boiling point**.

CHANGES OF STATE

Energy plays a key role in changes of state. Energy is required to increase the speed of particles of a substance. Therefore, energy is needed to change a substance from solid to liquid, or from liquid to gas. Each substance requires a specific amount of energy to make these changes. When a substance goes from gas to liquid or liquid to solid, it releases energy.

Temperature is a measure of the speed at which the particles in a substance move. If external heat is applied to a substance, it will generally increase its temperature. However, once the melting or boiling point is reached, the temperature will not continue to rise. This heat energy will be used by the substance to change its state. Until all the material has changed its state, its temperature will not increase. Once all the material has boiled or evaporated, adding more heat will begin to raise its temperature.

A scientist adds heat to a glass filled with cold water and ice cubes. The water is O° C (*the melting point of ice*) and the ice is at the same temperature. As heat is added to the glass, the heat is absorbed by the ice. The ice melts and changes to water. Until the rest of the ice changes into water, the water temperature will not increase. Once the ice has melted, added heat raises the temperature of the water.

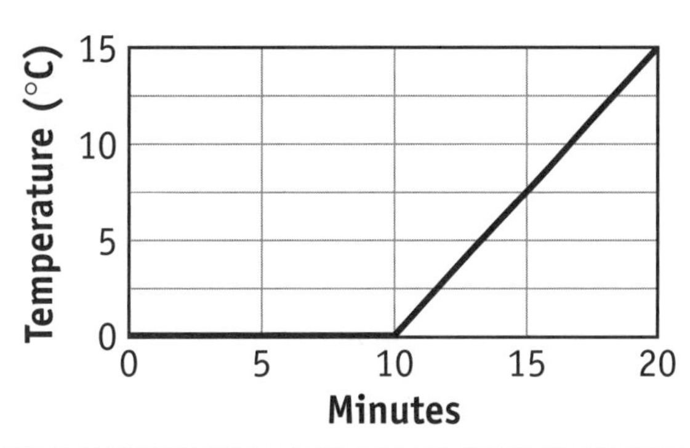

ACIDS, BASES, AND pH VALUES

One way to describe a mixture is by how acidic or basic it is. This is a common property of many solutions.

★ **Acids.** An **acid** is a special type of compound that forms excess hydrogen ions (**H^+**) when dissolved in water. These hydrogen ions combine with water molecules to form hydronium ions (**H_3O^+**). Acids have a sour or tart taste. Strong acids can dissolve many organic compounds. Some common acids are sulfuric acid and citric acid.

★ **Bases.** A **base** is a special type of compound that forms excess hydroxide ions (**OH^-**) in water. Bases have a bitter taste and a slippery feel. They are used to make soap and bleaching products. Some common bases are ammonia and lye. Because bases are often formed out of alkali metals, such as sodium or potassium, they are also called alkaline.

★ **Acid / Base Reactions.** When an acid combines with a base, they produce water and a salt. If you mix the acid hydrogen chloride with the base sodium hydroxide, you get water and table salt. These products are neutral. Think of combining acids and bases like mixing hot and cold water. A strong acid and strong base produce a neutral salt, while a strong acid and a weak base will produce an acidic salt.

★ **Determining pH Levels.** The **pH** level tells how acidic or basic a solution is. Scientists use a scale of **0 to 14** to indicate a solution's pH value. Each number below 7 is more acidic, while values above 7 are more basic. Each whole number below or above 7 is ten times greater. For example, a pH level of 9 is ten times ***more*** basic than a solution with a pH of 8. **Litmus paper** is often used in laboratory experiments to determine pH. An acid turns litmus paper red, and a base turns litmus paper blue.

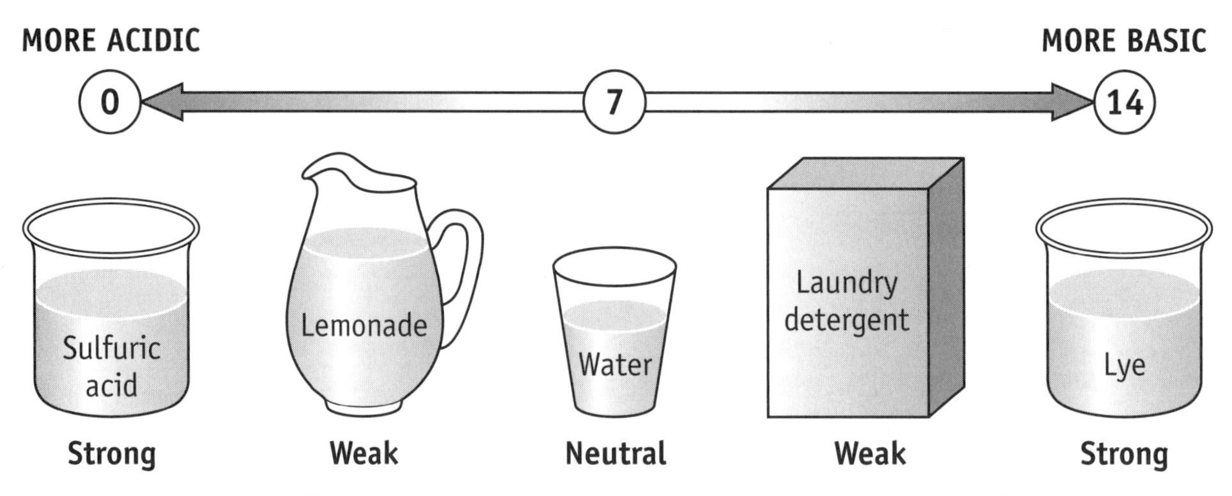

- **Bases** (like laundry detergents) have a pH value more than 7. ($H^+ > OH^-$)
- **Acids** (like lemonade) have a pH value less than 7. ($H^+ = OH^-$)
- **Neutral** solutions (like water) have a pH value of 7. ($H^+ < OH^-$)

APPLYING WHAT YOU HAVE LEARNED

✦ A student uses a pH meter to test an unknown solution. Its pH is 10. Is it an acid, base or salt? Predict how it will affect litmus paper and one other property.

ELECTRICITY AND CONDUCTIVITY

Another important property of any form of matter is whether or not it has the ability to conduct electricity. This is known as **conductivity**. Electricity is an electric charge that moves from atom to atom. Electrons act as the carriers of this electric charge. Temperature affects the conductivity of items. Materials are better able to conduct electricity when they are colder.

Substances with many free electrons are good conductors of electricity. Metals can conduct electricity because their atoms are so closely packed together that their electrons can move easily from one atom to another. Metalloids, also known as **semiconductors**, can conduct electricity under some conditions. **Superconductors** can conduct electricity extremely well. Nonmetals, like carbon and sulfur, do not conduct electricity. Conductivity also may change if the state of the material changes. Cold substances conduct electricity better than warmer ones. Ionically-bonded compounds like salt will not conduct electricity unless they are melted or dissolved in water. The melting or dissolving permits the ions to freely move about and carry an electrical charge. Solutions of acids, bases, and salts can all conduct electricity.

DENSITY

Another property of any form of matter is its density. The **density** of an object is determined by its mass divided by its volume. Density is often measured in grams per cubic centimeter (**g / cm^3**).

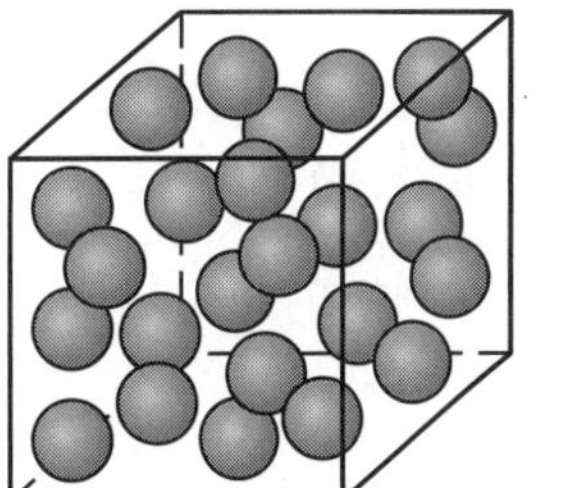
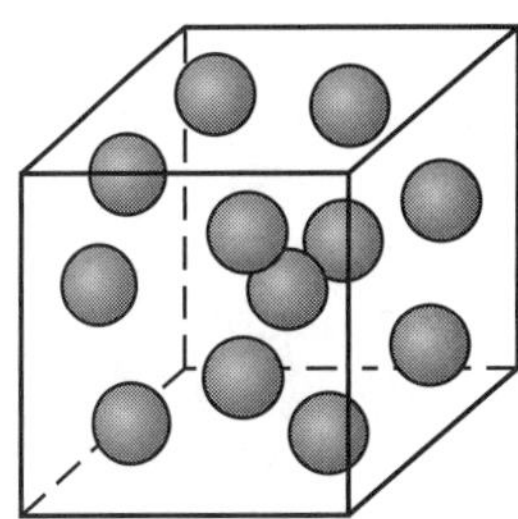

*These two boxes have the same **volume**. Assuming each ball has the same mass, the box with the greater number of balls has more mass per unit of volume.*

Determining the density of a substance, often helps scientists narrow down what the substance might be. The density of a sample of matter will change if the temperature or pressure applied to it changes; density also changes if the substance changes state from a solid to a liquid. A liquid usually has a greater volume than a solid because its particles are in greater motion and further apart. Water is an exception. Ice has greater volume than liquid water at 0°C.

Ability to Float. A solid that is less dense than a liquid will **float** in the liquid. For example, a penny will float in mercury because copper is less dense than mercury. However, the same penny will sink in water because copper has greater density than water. Similarly, a helium balloon floats in the air because helium is less dense than the atmosphere at Earth's surface.

APPLYING WHAT YOU HAVE LEARNED

✦ Explain why metals are good conductors of electricity.

✦ An object has a density of **1.29 g/cm^3** and a volume of **2 cm^3**. What is its mass?

WHAT YOU SHOULD KNOW

★ You should know how to describe the physical properties of substances. These include hardness, color, density, conductivity, and ductibility.

★ You should know how to read the Periodic Table of the Elements and understand that the periods and groups of elements in the Periodic Table are based on different arrangements of electrons. It also identifies repeating patterns of properties. Scientists classify elements as metals, metalloids, nonmetals, and noble gases.

★ You should know that acids have a pH value less than 7, that bases have a pH value greater than 7, and that a pH level of 7 is neutral.

★ You should know that metals and ionic solutions conduct electricity.

CHAPTER STUDY CARDS

Substances and Mixtures

★ **Pure Substance.** Any element or compound; a substance is the same throughout and has a fixed chemical composition.

★ **Mixture.** Two or more substances mixed together without being chemically combined. The substances can be separated without a chemical reaction.

Physical & Chemical Properties

★ **Physical Properties.** A substance's color, odor, density, melting point and boiling point, ductility, and conductivity.

★ **Chemical Properties.** The ability of a substance to react with other substances.

Periodic Table of the Elements

Chart of elements arranged by their atomic number — number of protons.

★ **Periods.** Begin with atoms containing one valence electron and end with atoms with a complete outer shell containing 8 electrons.

★ **Groups.** Members of a particular group have similar chemical and physical properties.

- Predicts how atoms of the different elements will combine.
- An element's position on the table will indicate many of its general properties.

Three States of Matter

- ★ **Solids.** Atoms, ions, and molecules are locked into fixed positions, giving the substance a fixed volume and shape.
- ★ **Liquids.** When energy is transferred to a substance, its particles start to move around, melting the solid into a liquid. The temperature at which it turns into a liquid is its **melting point**. Liquids have volume but no fixed shape.
- ★ **Gas.** If more heat is applied to a liquid, its particles will move more rapidly, breaking all connections and turning into a gas. The temperature at which liquid turns into a gas is its **boiling point.**

Acids and Bases

- ★ **Acid.** A special type of compound that forms hydrogen ions when dissolved in water. Acids are tart or sour.
- ★ **Base.** A special type of compound that forms hydroxide ions in water. Bases are slippery and taste bitter.
- ★ **pH Levels.** Scientists use a scale of 0 to 14 to indicate pH values:
  - **Bases** have a pH value of more than 7.
  - **Acids** have a pH value of less than 7.
  - **Neutral** solutions have a pH value of 7.
- ★ **Acid/Base Reactions.** An acid combines with a base to produce water and salt.

CHECKING YOUR UNDERSTANDING

1 The Periodic Table of the Elements can be used by scientists

A. to find out the main uses of each element

B. to predict how atoms of different elements will combine

C. to identify all of an element's physical and chemical properties

D. to determine the differences between ionic and covalent bonding

- ◆ Examine the Question
- ◆ Recall What You Know
- ◆ Apply What You Know

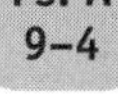

This question examines your understanding of the Periodic Table. Recall that the Periodic Table arranges elements into periods and groups based on their electron configurations. Since the table shows how many valence electrons are in an atom's outermost energy level, it can also be used to identify how atoms of different elements will combine.

Now try answering some questions on your own about the properties of matter and the Periodic Table.

2 What are two properties of most nonmetals?

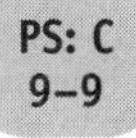

A. They form negative ions and do not conduct electricity.

B. They form negative ions and conduct electricity well.

C. They form positive ions and do not conduct electricity.

D. They form positive ions and conduct electricity well.

Use the Periodic Table to answer questions 3 to 7

PARTIAL PERIODIC TABLE OF THE ELEMENTS

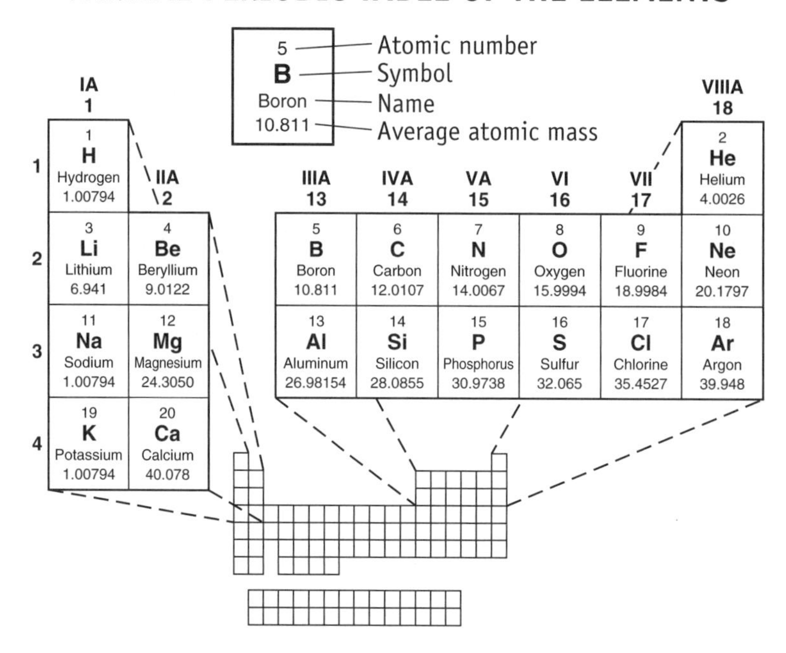

3 Based on the Table, which of these elements probably has physical and chemical properties most similar to boron (B)?

PS: C 9–4

A. magnesium (Mg)
B. aluminum (Al)
C. neon (Ne)
D. chlorine (Cl)

4 Which is a property of the noble gases in Group 18?

PS: C 9–9

A. malleability
B. brittleness
C. high electrical conductivity
D. unlikely to react with other elements

5 The elements in the Periodic Table are arranged in order of increasing

PS: C 9–4

A. atomic number
B. hardness in solid form
C. atomic radius
D. neutron number

6 How many valence electrons does oxygen (O) have in its outer most energy level?

PS: A 9–4

A. 4
B. 6
C. 8
D. 16

7 Which of the following elements is a nonmetal?

PS: C 9–9

A. fluorine (F)
B. calcium (Ca)
C. magnesium (Mg)
D. sodium (Na)

8 The major difference between an element and a compound is that in an element

A. all the atoms are the same
B. there are atoms of different types
C. molecules can be chemically combined
D. two or more substances are mixed together

◆ Examine the Question
◆ Recall What You Know
◆ Apply What You Know

PS: B
9–9

9 The density of water is 1 g/cm^3. Substances whose density is less than the density of water will float on the surface of water. Which of the following objects will float on water?

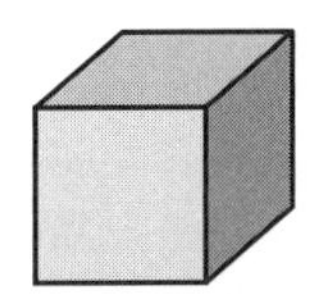

A. a cube with a mass of 45 g, and a volume of 10 cm^3

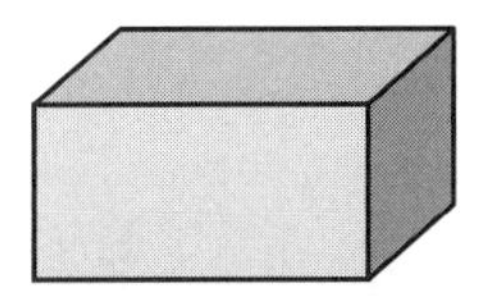

C. a box with a mass of 55 g, and a volume of 100 cm^3

PS: C
9–9

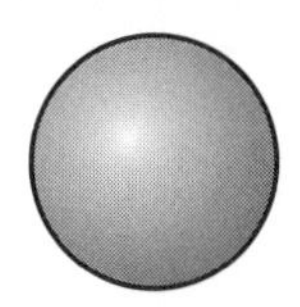

B. a sphere with a mass of 45 g, and a volume of 15 cm^3

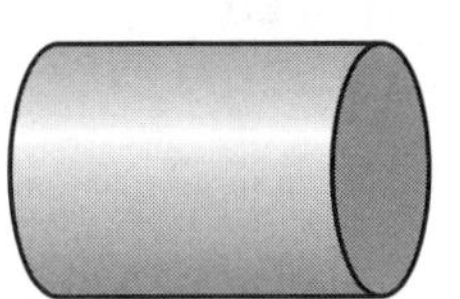

D. a cylinder with a mass of 120 g, and a volume of 60 cm^3

10 Which statement is true concerning a substance with a pH level of 13?

A. The substance is an acid.
B. The substance is a strong base.
C. The substance is a weak base.
D. The substance is neutral.

PS: C
9–8

11 Two substances are physically blended together without chemically reacting. They retain their original chemical and physical properties. What is this combination of substances called?

A. an element
B. a molecule
C. a compound
D. a mixture

PS: C
9–9

12 The metal silver allows conduction of an electric current because its electrons are

A. not negatively charged
B. covalently shared
C. able to move between atoms
D. drawn towards neutrons

PS: C
9–10

13 Different types of metals have been highly valued from earliest times. Today, several metals are still prized for use in such items as fine jewelry. Name two physical properties metals have that make it favored by jewelers. (*2 points*)

PS: C
9–9

CHAPTER 13

FORCE AND MOTION

In this chapter, you will learn about force and motion. In particular, you will learn about Newton's three laws of motion.

MAJOR IDEAS

A. **Motion** refers to how an object changes its position over time. **Velocity** is both the speed and direction an object moves. **Acceleration** is any change in an object's velocity. **Momentum** is the mass of an object times its velocity.

B. **Newton's First Law:** An object remains at rest or maintains a constant speed and direction, unless an unbalanced force acts on it.

C. **Newton's Second Law:** The net force acting on an object equals the object's mass times its acceleration. A force of one newton (1N) will accelerate the motion of one kilogram by one meter per second each second ($1N = 1 \text{ kg.} \bullet 1 \text{ m / sec}^2$)

D. **Newton's Third Law:** For every action, there is an equal and opposite reaction.

E. **Friction** occurs when objects rub against each other; it opposes motion.

MOTION, DISTANCE, AND VELOCITY

Sir Isaac Newton revolutionized how scientists think about the natural world. One of his discoveries was the law of gravity — the force that explains not only why things fall to Earth, but how the planets move in our solar system. Newton also revolutionized how scientists think about motion. He showed that motion throughout the universe always follows the same rules. These rules are precise enough to make predictions that can be checked by careful measurement.

What exactly is motion? **Motion** is any change in an object's position over time. Motion is a measurable quantity. The difference in position, the time the motion took, and the speed of the motion can all be measured.

Scientists distinguish between displacement and distance. **Displacement** refers to an object's net change in position. It is the shortest distance between the object's starting position and location at the end of the motion. **Distance** is the total amount of ground covered by the object in motion. It is the whole length moved by the object, and takes into account all movements back and forth, side to side, and so forth. Consider the object moving from point **A** to **B** in the diagram. Its displacement from its starting point is 5m, but it has moved a total distance of 7m.

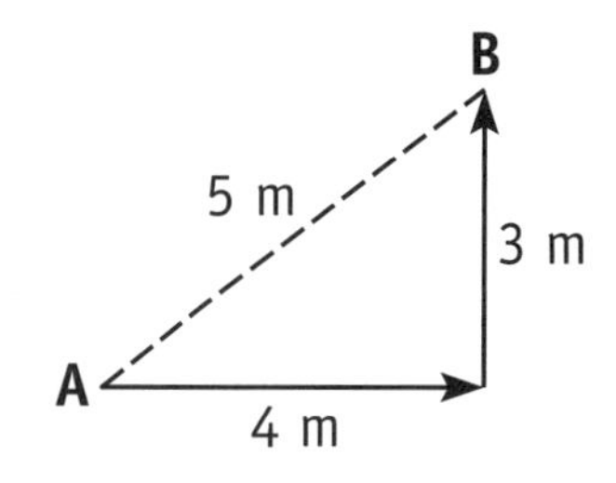

Velocity refers to both the **speed** at which an object moves *and* the **direction** that it moves in.

$$\textbf{Velocity} = \frac{\text{Distance traveled}}{\text{change in time}}$$

Velocity is affected by the "frame of reference" of the observer. If the object moves in one direction at 10 m/s and the observer moves at the same rate in the opposite direction, the speed of the object is 20m/s with respect to the observer. If they both move in the same direction at the same speed, then the object will seem to be at rest.

NEWTON'S FIRST LAW OF MOTION

Aristotle and other ancient philosophers had believed that all objects were naturally at rest. Therefore, it took energy for an object to keep moving. They believed that if a ball was rolling, it would stop rolling unless additional force was applied to it. Newton took a very different view. He believed that friction slowed down most movements we observed. Without friction or some other opposing force, Newton said a moving object would continue moving forever in the same direction at the same speed.

Newton's First Law states that an object in motion will stay in motion unless some outside force acts on it.

This idea became his first law of motion, also known as the **Law of Inertia**: an object at rest will stay at rest, or if in motion it will continue to move in a straight line at the same speed, unless some outside force acts on it. **Inertia** is the property of matter that resists changes in motion.

NEWTON'S SECOND LAW OF MOTION

Sometimes an object will continue to move at the same speed for a long period of time. For example, your parents may drive along the highway using their car's "cruise control." The family car moves at 50 miles per hour, and continues to move for two hours at that speed. This is known as **uniform motion**.

Acceleration. Sometimes, however, your parents may speed up or slow down the car. Scientists refer to such **changes** in velocity, whether becoming faster or slower, as **acceleration**. Even a change in direction at the same speed is considered as acceleration by scientists. Acceleration, like velocity, can be measured. It is the rate at which velocity changes.

> Thinking about acceleration can be confusing at first. Imagine your uncle is driving down a street at 25 miles per hour. He enters the highway. Over the next ten minutes, he increases his speed to 55 miles per hour. What was the *change* in his velocity? Over ten minutes, he increased his velocity by 30 miles per hour. What was rate of change — the average change in his velocity each minute? If he increased his speed by 30 miles per hour in ten minutes, then he increased his speed *each minute* by 3 miles per hour. His acceleration was 3 miles per hour per minute.

Scientists generally measure acceleration by changes in meters per second per second, or m / s^2. Don't forget that scientists use acceleration to refer to slowing down as well as speeding up. Because velocity refers to direction as well as speed, scientists also use the term "acceleration" to describe a change in direction even when the speed remains the same.

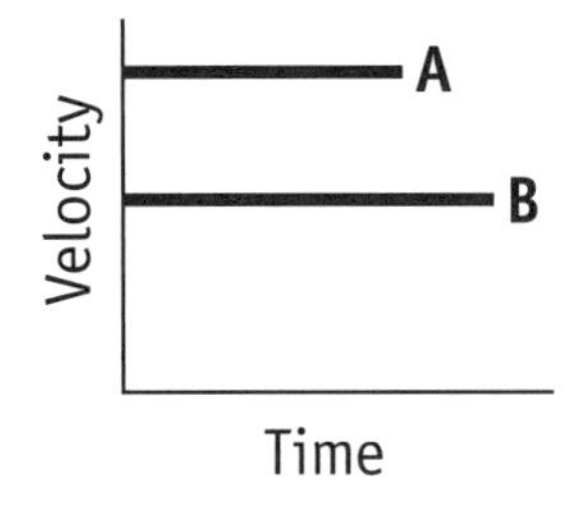

A and **B** move at constant velocity; **A** travels faster than **B**. Both have zero acceleration.

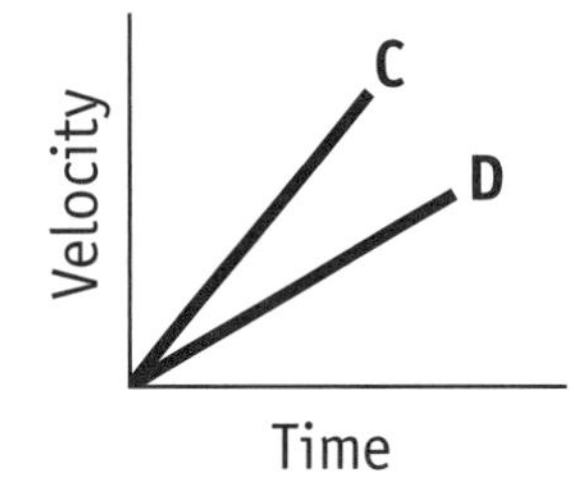

Both **C** and **D** accelerate uniformly but **C** has a greater acceleration.

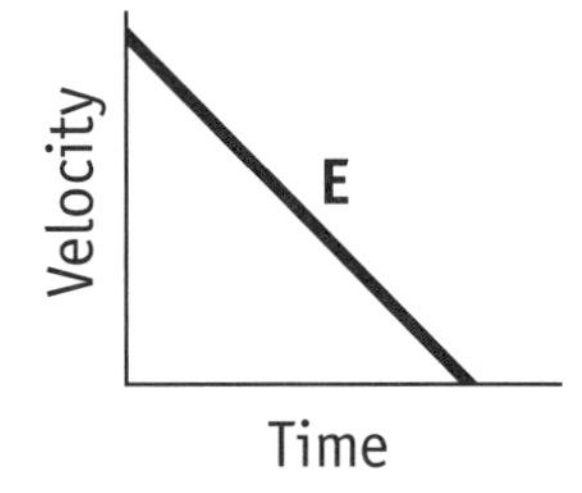

E is accelerating negatively (slowing down).

APPLYING WHAT YOU HAVE LEARNED

- ✦ Dora's parents are driving at 20 miles per hour. Her mother pushed the car's gas pedal increasing its speed. If the car takes 10 minutes to increase its speed up to 40 miles per hour, what is the rate of acceleration in miles per hour per minute?
- ✦ Draw a line graph to show the acceleration of Dora's parents' car. Label one axis "Time" and the other axis "Velocity."

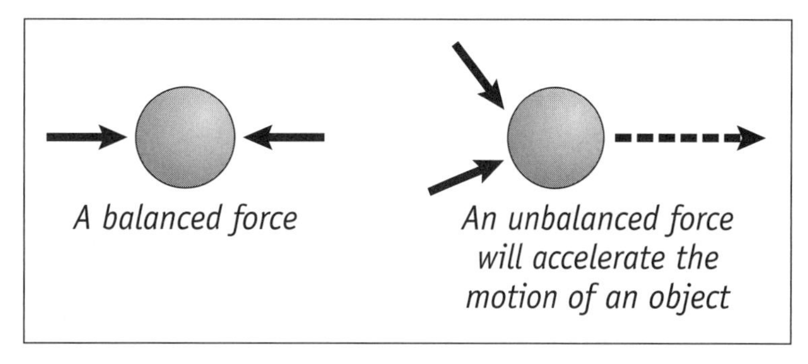

In the First Law of Motion, Newton had concluded that an object on its own without additional force would not slow down, speed up, or stop. It would simply stay at rest if it was already at rest, or keep moving at the same velocity if it was already in motion. The **Second Law of Motion** is a logical extension of Newton's First Law. This law states that some **unbalanced**, outside force must be required for an object to change its motion (*or state of rest*). An "unbalanced" force means that if several forces are acting on an object, they are not perfectly balanced — the overall, or "net" effect of these forces will be to push the object in one direction.

In other words, to increase the speed of an object in motion, you have to give it a "push" in the same direction. To slow it down, you have to give it a "push" in the opposite direction. To get it moving from a state of rest, you have to give it a push" in the direction you want it to move.

How great must this "push" or force be to change the velocity of the object? Newton's second law also answers this question:

★ If the mass of the object is greater, the force must be *greater*.

★ For the acceleration to be greater, the force must be *greater*.

Thus, the force required must be ***proportional*** to **both** the mass of the object and the change in acceleration:

Force = mass • acceleration

Scientists use "newtons" as the units to measure force. One newton (1N) is the amount of force needed to accelerate the motion of one kilogram by one meter per second per second. **1N = 1 kg • 1 m / sec^2**

Suppose a 5 kg asteroid in frictionless outer space moves at the velocity of 100 m/s. How many newtons are needed to increase its velocity from 100 m/s to 200 m/s in one second?

SOLUTION

- The mass of the asteroid is 5 kg. Since the asteroid increases its velocity from 100 m/s to 200 m/s in one second, its acceleration is 100 m/s^2.
- Force = m • a
- Force = 5 • 100
- Force = 500 N

Weight vs. Mass. The weight of an object is a direct measure of Earth's gravitational attraction. It is not the same as mass, which is a measure of the amount of matter in an object. Weight is proportional to mass, however, since an object with greater mass will weigh more on Earth's surface. Scientists measure mass in grams and weight in newtons.

Constant Rate of Acceleration in Free Fall. In ancient times, people believed that heavier objects fell faster than light objects. Air resistance does, in fact, make some things fall more slowly than others. For example, a bowling ball will fall more quickly than a feather. Without air, however, these objects would fall at the same speed. If you dropped a bowling ball and a feather from the same distance in a vacuum, they would actually land at the same time.

Newton's Second Law of Motion explains why this is so. Recall that the force of gravity between two objects is proportional to their masses. That means that the force of gravity attracting the bowling ball is greater than the force attracting the feather. Even though the force of gravity is greater for the bowling ball than for the feather, Newton's Second Law shows that a greater force is needed to accelerate its larger mass. In fact, the larger gravitational force and greater force needed exactly cancel each other out. As a result, without air resistance all objects would fall to Earth at the exact same rate of acceleration, regardless of size: 9.8 meters/second^2.

APPLYING WHAT YOU HAVE LEARNED

✦ The same force is applied to two objects in frictionless outer space. One object is 500 kg and the other is 1,000 kg. Which object will accelerate faster? How much faster? Explain your answer.

NEWTON'S THIRD LAW OF MOTION

Newton's Third Law of Motion states that whenever one object exerts a force on a second object, the second object exerts an ***equal and opposing*** force on the first object. Simply stated, for every action, there is an equal and opposite reaction.

Action/Reaction Forces. Newton's Third Law of Motion states that forces always occur in pairs: an "action" force and "reaction" force. For example, a book placed on a table pushes down on the table with a force equal to the book's weight. This is the "action" force. The table pushes up on the book with an equal and opposite "reaction" force. The net effect of these two forces is that the book remains at rest. Newton's Third Law of Motion explains how jet engines and rockets work. As hot gases are pushed out of the back of the jet or rocket ("action force"), the vehicle is pushed forward ("reaction force").

MOMENTUM

Newton's Third Law of Motion requires an understanding of momentum. Think about the difference between being hit by a falling snowflake and a speeding railroad train. The snowflake is light and falls slowly. The train has a large mass and a faster velocity. Its greater momentum will have a greater impact. Momentum reflects both the mass and speed of an object. The momentum of an object is equal to its mass times its velocity. The equation for the momentum of an object is **p = mv**, where **p** is the momentum, **m** is the mass, and **v** is the velocity.

Newton's Third Law of Motion is based on the principle of the **Conservation of Momentum**. The momentum of an *action force* will equal the momentum of the *reaction force*. When a bullet is shot from a rifle, the small mass of the bullet flies off in one direction with great velocity. At the same time Newton's action-reaction law states that the rifle will recoil in the opposite direction. The rifle's larger mass and slower velocity will have a momentum equal in amount but opposite in direction to that of the bullet.

APPLYING WHAT YOU HAVE LEARNED

- ✦ What is the momentum of an object at rest?
- ✦ A man slips and falls on the floor. What is the action force? The reaction force?

FRICTION

Friction is a force that resists or opposes the motion of objects in contact. All surfaces are irregular to some degree. Friction results from the rubbing of these irregular surfaces. There are three common types of friction:

- ★ **Static friction** is the friction that exists between the surfaces of objects at rest. Static friction must be overcome for an object to begin motion.
- ★ **Sliding friction** is the rubbing of surfaces of objects already in motion. Such friction will slow down and stop motion unless some additional force is applied to overcome it.
- ★ **Rolling friction** is the friction between a round object, such as a wheel, and a flat surface. Rolling friction is generally less than sliding friction.

Friction slows down and eventually stops the movement of objects on Earth unless additional force overcomes it. The amount of frictional force between two objects depends on the roughness of their surfaces and how much force pushes them together. Friction is greater as velocity increases because there is more rubbing. Smooth surfaces have less friction. Lubricants, such as oil, are often applied to machine parts to reduce friction.

A car with new tires stops more easily than a car with old ones. Worn-out tires are smoother and don't grip the road as well because of less friction. Car brakes equally depend on friction to stop a car. Part of the brake rubs on the brake lining. When wet, brakes do not work as well. The water smooths the surface of the brake lining, reducing friction.

A block of wood on an inclined plane experiences a frictional force opposing its motion. The larger the angle of the inclined plane, the smaller the frictional force and the more likely the block will slide down the incline. Frictional force such as air resistance slows down airplanes in flight but also slows down parachutes as they descend toward the ground. To decrease friction, designs of automobiles and airplanes are developed to minimize air resistance.

APPLYING WHAT YOU HAVE LEARNED

- ✦ Explain how friction is essential for a car to work properly.
- ✦ Explain how engineers overcome unwanted friction.

WHAT YOU SHOULD KNOW

- ★ You should know that velocity is the speed and direction in which an object moves. **Acceleration** refers to changes in **velocity**. Newton's three laws of motion state: (1) an object remains at rest or in motion at the same speed and direction unless an unbalanced force acts upon it; (2) the net force acting on an object equals its mass times acceleration; and (3) for every action, there is an equal and opposite reaction.
- ★ Friction results from the rubbing of uneven surfaces and opposes motion.

CHAPTER STUDY CARDS

Key Motion Concepts

- ★ **Displacement.** Change in an object's position.
- ★ **Distance.** The total length moved by the object (in meters).
- ★ **Velocity.** The rate and direction an object moves.
- ★ **Acceleration.** The change in an object's velocity (meters / second / second)
- ★ **Momentum.** Mass times velocity.
- ★ **Friction.** Force from rubbing surfaces that opposes motion.

Newton's Laws of Motion

- ★ **First Law of Motion.** (Law of Inertia). An object at rest stays at rest, and an object in motion stays in motion at the same velocity (*speed and direction*) unless an unbalanced force acts upon it.
- ★ **Second Law of Motion.** (Force on Accelerating Body). The net force acting on a body equals its mass times acceleration: F net = m • a
- ★ **Third Law of Motion.** (Action/Reaction Law). For every action, there is an equal and opposite reaction.

CHECKING YOUR UNDERSTANDING

Use the graphs to answer the following question

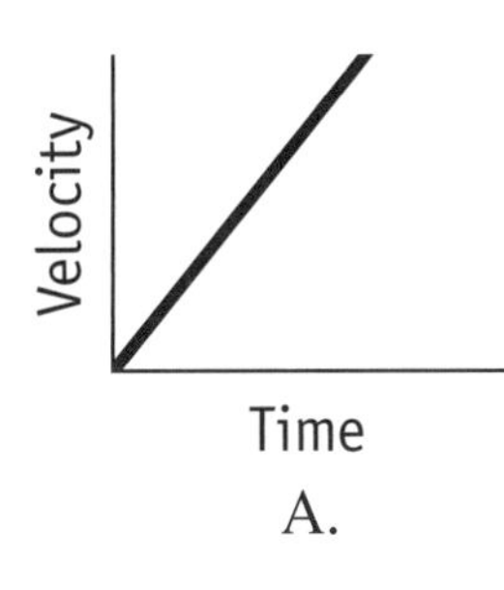

A.

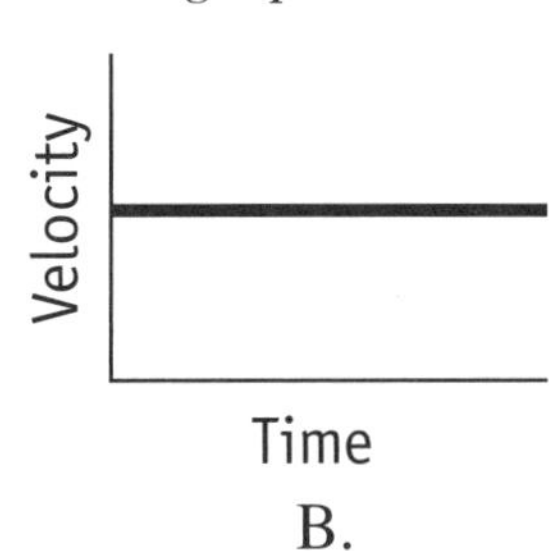

B.

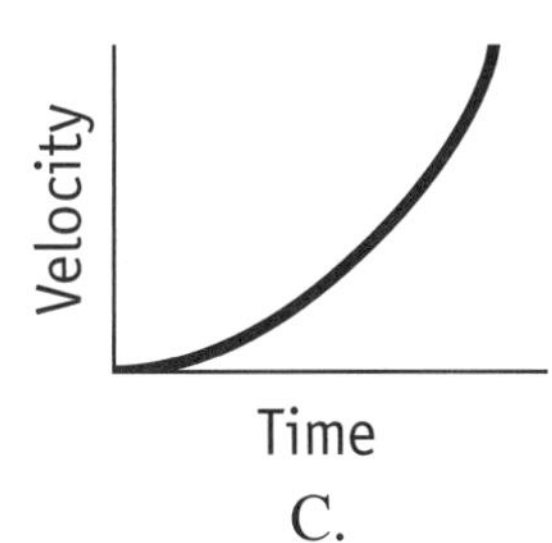

C.

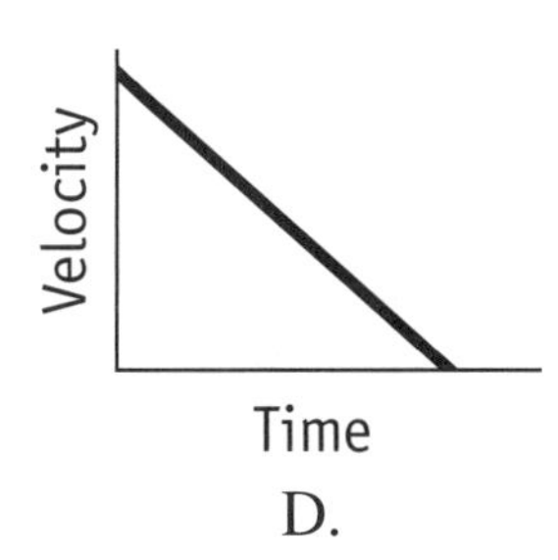

D.

1 Which graph represents the motion of an object on a frictionless surface on which the net force is zero?

A. A
B. B
C. C
D. D

- Examine the Question
- Recall What You Know
- Apply What You Know

PS: D 9–22

This question asks you to examine a series of graphs and to relate them to Newton's laws. A net force of zero means there is no force acting on the object. Recall that if there is no force acting on an object, it will stay at rest or in motion without change. Then you must apply this knowledge to the graphs. Since all of them show some velocity, the object is in motion at some point. Choices A, C and D show the speed changing over time, so they must be incorrect. Choice B shows a constant speed, and is therefore the correct answer.

Now try answering some questions on your own about force and motion.

2 If you push on the wall with a force of 22 newtons, what is the force being exerted by the wall on your hand?

A. 0N
B. 44N
C. 22N
D. 11N

PS: D 9–24

3 A 4.0-kilogram rock and a 1.0-kilogram stone fall freely from rest at a height of 100 meters. After they fall for 2.0 seconds, the ratio of the rock's speed to the stone's speed is

A. 1:1
B. 2:1
C. 1:2
D. 4:1

PS: D 9–23

4 Which of the following does *not* indicate velocity?

A. 14 feet per second per second
B. 40 miles per hour towards town
C. 80 miles per hour towards New York
D. 28 miles per second towards Ohio

♦ Examine the Question
♦ Recall What You Know
♦ Apply What You Know

PS: D
9–21

5

A baseball rolling in the outfield comes to a stop. Two baseball players give an explanation:

- **Shortstop: "A baseball's natural state is at rest, so the ball slowed down to reach its natural state."**
- **Catcher: "The ball slowed down and stopped because forces acted upon the baseball to make it come to a stop."**

According to Newton's Laws of Motion, which of the ball players is correct?

A. the shortstop
B. the catcher
C. both are correct
D. neither is correct

PS: D
9–25

6 Meagan applied a constant force on a small box, causing the box to accelerate. After some period of time, the box stopped accelerating. What conclusion can Meagan reach from this experience?

A. The mass of the box has increased greatly.
B. The force of gravity on the object has increased.
C. The box is experiencing some kind of frictional force.
D. The box has reached its maximum potential momentum.

PS: D
9–25

7 If the sum of all the forces acting on a moving object is 0, the object will

A. slow down and stop
B. change the direction of its motion
C. accelerate uniformly
D. continue moving with constant velocity

PS: D
9–22

8 The diagram to the right illustrates a large net force being constantly applied to a basketball. Students are measuring the motion of the ball as part of an experiment they are conducting. Identify two measurements that the students should take and record to describe the motion of the basketball. (*2 points*)

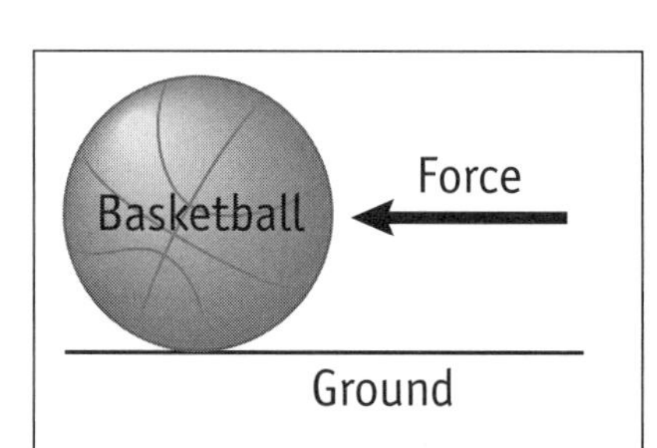

PS: D
9–21

THE NATURE OF ENERGY

In this chapter, you will learn about the nature of energy.

MAJOR IDEAS

A. **Energy** is the ability to do work. Energy can take different forms, but it is always conserved.

B. An object's **kinetic energy** depends on its mass and speed. An object's **gravitational potential energy** depends on its weight and height.

C. **Thermal energy** exists in the random motion of atoms and molecules. It may be transferred by conduction, convection, or radiation.

D. **Nuclear reactions** convert a small amount of matter into large amounts of energy.

E. **Waves** have energy and can transfer energy. The properties of a wave depend on the material (*medium*) through which it travels.

F. **Electromagnetic radiation**, including light, is a form of energy that acts as a wave. It can pass through a vacuum.

TYPES OF ENERGY

As you know, **force** must be applied to an object to change its motion. **Work** is the application of force over distance. **Energy** is the ability to do work. Energy is measured in **joules**. There are many types of energy, including:

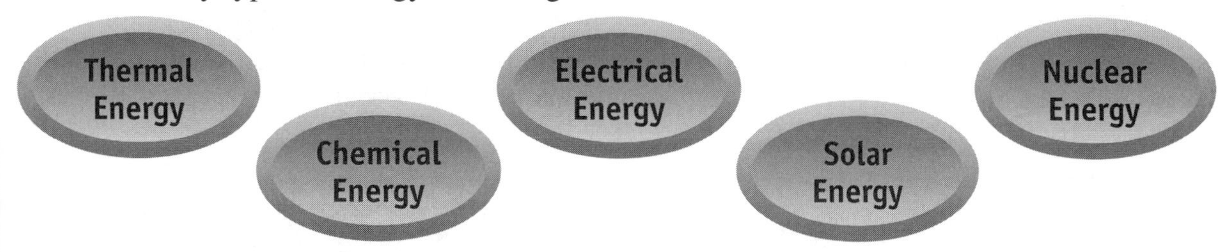

Most forms of energy can be classified as either kinetic energy or potential energy. Let's examine each to see how they differ.

KINETIC ENERGY

Any moving object is able to do work: therefore it has energy. This energy of motion is known as **kinetic energy**. When you walk or run your body exhibits kinetic energy. The amount of kinetic energy a moving object has depends on two factors: its mass (m) and velocity (v).

Kinetic Energy = 1/2 (mass) (velocity^2)
$KE = 1/2\ mv^2$

As you can see, the kinetic energy of a moving object is directly proportional to the mass of the object and the square of its speed. The greater its speed, the greater its kinetic energy; the greater its mass, the greater its kinetic energy. The speed of movement will have more of an impact on the total amount of kinetic energy than the mass. If the speed of a moving object doubles, its kinetic energy will increase fourfold. For example, a car moving at 20 miles per hour will have four times the kinetic energy of the same car moving at only 10 miles per hour.

Science Photo Library

Because this ball is moving, it has kinetic energy and can use this energy to move other objects.

Thermal Energy. Some types of non-mechanical energy, like thermal energy (*heat*), are also forms of kinetic energy. The **kinetic theory** explains that what we feel as heat is actually caused by the random motion and vibration of atoms and molecules in substances. The heat generated by this movement is therefore a form of kinetic energy. **Temperature** measures the speed of movement of these atoms and molecules: it is the average kinetic energy of the particles in an object. An increase in temperature represents an increase in molecular motion. The higher the temperature of an object, the greater the kinetic energy of its atoms and molecules. **Absolute zero** (– 273.16° C) is the lowest possible temperature because it represents the point at which all molecular motion stops.

APPLYING WHAT YOU HAVE LEARNED

✦ Use the kinetic theory to explain why the temperature of ice stays constant as it melts.

POTENTIAL ENERGY

Potential energy is the energy that an object is able to store because of its position or condition. Think of a metal spring. As the spring is pushed down, potential energy is stored in its coils. Once you let go of the spring, it bounces back up, converting its potential energy into kinetic energy.

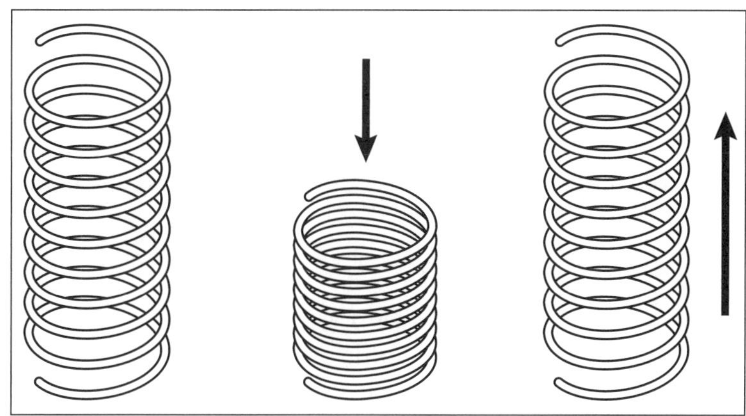

The spring stores potential energy when pushed down. When it is released, it converts its potential energy to kinetic energy (motion).

GRAVITATIONAL POTENTIAL ENERGY

Gravitational potential energy is one form of potential energy. It is the type of energy that an object has because of the work that was done against the force of gravity to move it to its present position. This position on Earth's surface gives the object the ability to exert force and do work. For example, if a truck is pushed to the top of a hill, it can roll down the hill and pull an object behind it. Energy was "stored" in the truck when it was pushed up the hill. The amount of potential energy stored in the truck will depend on its weight and its height.

Remember that an object's weight on Earth's surface is equal to its mass times the rate of acceleration of free-falling objects. Scientists use the following equation to calculate an object's gravitational potential energy:

Gravitational P.E. = m g h

Where, **m** = mass of the object
g = free-fall acceleration rate
h = height of the object

Gravitational potential energy will therefore be greater if the mass of an object is greater. It also increases the higher an object moves above Earth's surface. Thus, a ball suspended 100 feet high will have twice the gravitational potential energy of a ball of the same weight that is suspended only 50 feet high.

CHEMICAL ENERGY

Chemical energy is another form of potential energy. Some types of molecules store energy in their covalent bonds. When these bonds are broken, their energy is released.

OTHER FORMS OF ENERGY

ELECTRICITY

Another form of energy is electricity. Electricity is created by the movement of electrons. These electrons carry negative electrical charges, which can travel through substances with free electrons and can even move from one substance to another.

NUCLEAR ENERGY

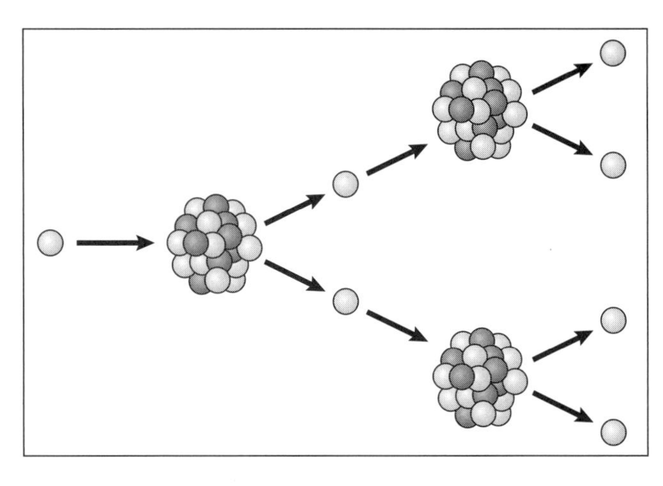

Nuclear energy is another important form of energy. As you may recall, atoms with large nuclei are often unstable. In the 1930s, this led to an important discovery. Scientists learned they could bombard some large atoms with neutrons to split their nuclei apart, in a process known as **nuclear fission**. When these large nuclei split apart, they release free neutrons and energy. Small amounts of matter are converted into immense quantities of energy. The free neutrons in turn bombard other nuclei, causing them also to split apart. This creates even more energy, and even more free neutrons — leading to a **chain reaction**.

Scientists used nuclear fission to develop the first atomic bomb — an uncontrolled **nuclear reaction**. Nuclear chain reactions can be controlled by the use of control rods, which absorb extra neutrons. These safer nuclear reactions are used today in nuclear reactors to generate electrical power.

Nuclear fusion is the opposite of fission. In fusion, the nuclei of smaller atoms are joined together, releasing stored nuclear energy. The sun, for example, fuses together the nuclei of hydrogen atoms into helium to produce its energy.

An atom bomb was dropped on Japan during World War II.

APPLYING WHAT YOU HAVE LEARNED

- Which has greater gravitational potential energy: a weight of 20 newtons that is 5 meters high, or a weight of 5 newtons that is 10 meters high? Explain.
- How do nuclear fission and fusion differ? How are they similar?

THE TRANSFER OF ENERGY

One of the qualities of energy is its ability to move. This gives energy the ability to move through a substance or even to transfer from one substance to another.

THERMAL ENERGY

Thermal energy can spread from one object to another by either ***conduction***, ***convection***, or ***radiation***. This transfer of energy is known as **heat**. Heat always transfers from an object with a higher temperature to one with a lower temperature.

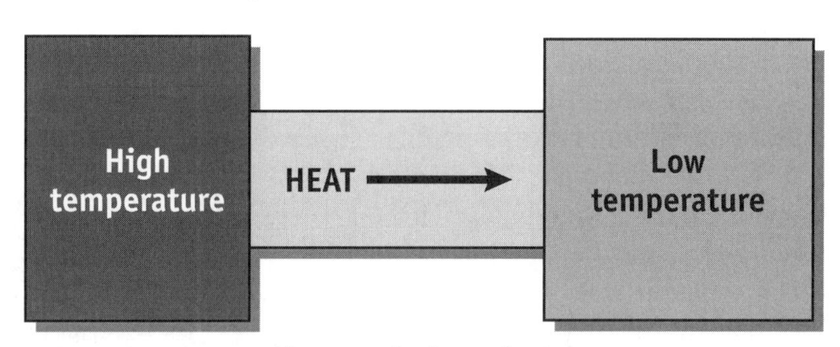

Heat transfer by conduction

Conduction. Conduction is the transfer of thermal energy within an object or between two objects in direct contact. When one object touches another, some of its vibrating and moving atoms and molecules collide with the atoms and molecules of the second object, causing them to move more rapidly. Some substances are better conductors of heat than others. Metals, such as aluminum or copper, are good conductors of heat, while items made of wood are poor conductors of heat.

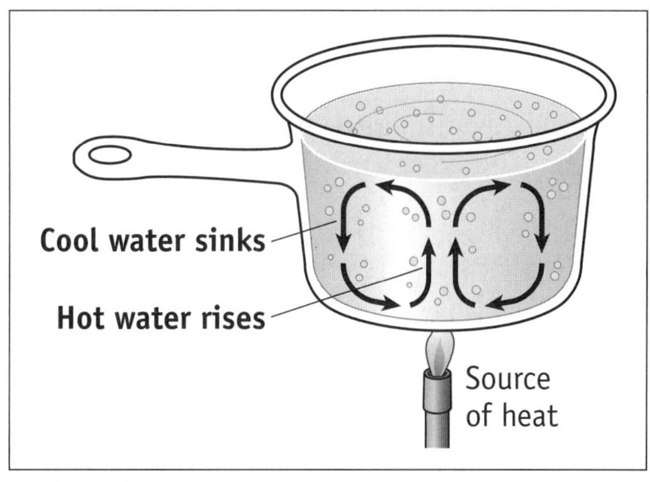

A pot of water over a fire creates a convention current.

Convection. When a **fluid** (*a liquid or gas*) is heated, its atoms and molecules become more active and it expands. Because there is more space between its particles, it becomes lighter. Hotter fluids therefore rise. As the heated fluid rises, the surrounding cooler, denser fluid takes its place. The heated fluid moves away from the source of heat and gradually cools. As it cools, it contracts and sinks down again, while the newly heated fluid expands and rises. This creates a circular movement known as a **convection current**. You may recall that you studied convection currents when you studied Earth's mantle, ocean currents, and the movement of air in Earth's atmosphere.

Radiation. Unlike conduction and convection, radiation does not require contact between substances. It is the transfer of energy through waves. Waves of energy radiate from a fire. These waves can be absorbed by other objects when they cause their atoms and molecules to vibrate more rapidly. We feel this energy as heat. Unlike conduction and convection, radiation can pass through a vacuum. Sunlight is a form of radiation which passes through space.

THE TRANSFORMATION AND CONSERVATION OF ENERGY

Energy also has the ability to change its form. For example, potential energy can change into kinetic energy, and kinetic energy can change into potential energy.

Think about a roller coaster ride. The roller coaster car is brought by a conveyer belt to the top of a hill. It now has stored gravitational potential energy. As it starts downward the roller coaster car's gravitational potential energy decreases, but its kinetic energy increases. This kinetic energy then pushes the roller coaster car up the next incline, increasing its gravitational potential energy again, while its kinetic energy decreases as it slows down. The sum of its kinetic energy and gravitational potential energy equals its total **mechanical energy**.

As a roller coaster car loses height, it gains speed, Potential energy is converted to kinetic energy. As it gains height, the car loses speed: the kinetic energy is transformed into potential energy.

Endothermic and Exothermic Chemical Reactions. Other kinds of energy, besides mechanical energy, can also change their form. For example, chemical energy often converts into thermal energy (*heat*) when a chemical reaction occurs; other chemical reactions may absorb energy. Chemical reactions that require energy are known as **endothermic reactions**. For example, in photosynthesis, plants take light energy from the sun and convert this into chemical energy. Chemical reactions that release energy are **exothermic reactions**. Exothermic reactions, like the burning of wood, often transform chemical energy into thermal energy that is released into the environment.

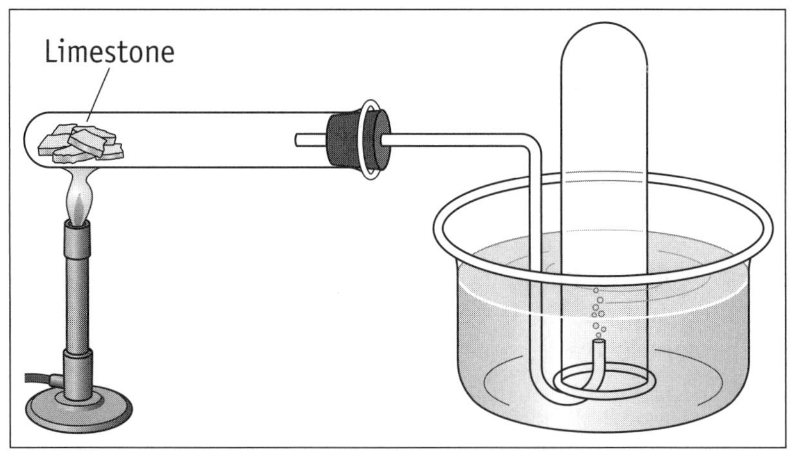

Endothermic reactions usually require heat. Limestone can break down into lime and carbon dioxide gas, but only if the limestone is heated.

Machines. Power-driven machines make use of the ability of energy to change its form. The internal combustion engine found in automobiles uses the chemical energy stored in fossil fuels (*gasoline*). It releases this energy by igniting the fuel, causing an explosion of gas and heat. The expansion caused by molecular kinetic energy is then used by the engine to force pistons to go up and down. The motion of the pistons rising and falling is carried to the wheels. In this example, the engine has turned chemical energy into thermal energy and then into mechanical energy.

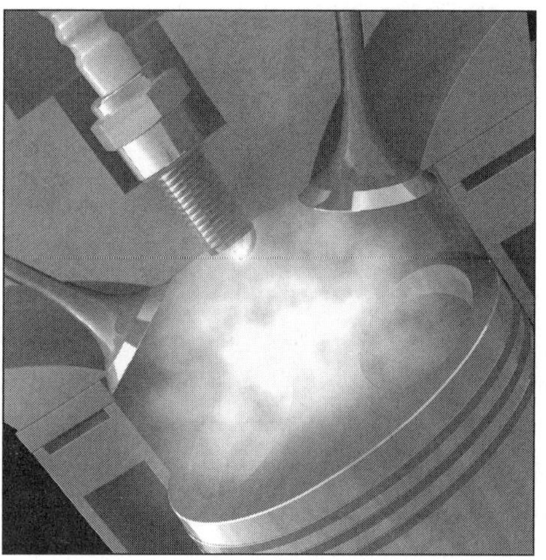
Science Photo Library

APPLYING WHAT YOU HAVE LEARNED

✦ List two examples from everyday life in which energy changes its form.

THE CONSERVATION OF ENERGY

Although energy can change from one form to another, it cannot be created or destroyed. This principle is known as the **conservation of energy**.

Sometimes it may seem that energy is being lost. For example, when a pendulum swings back and forth, its gravitational potential energy is converted to kinetic energy and back again. Each pendulum swing, however, is slightly shorter. Eventually, the pendulum stops.

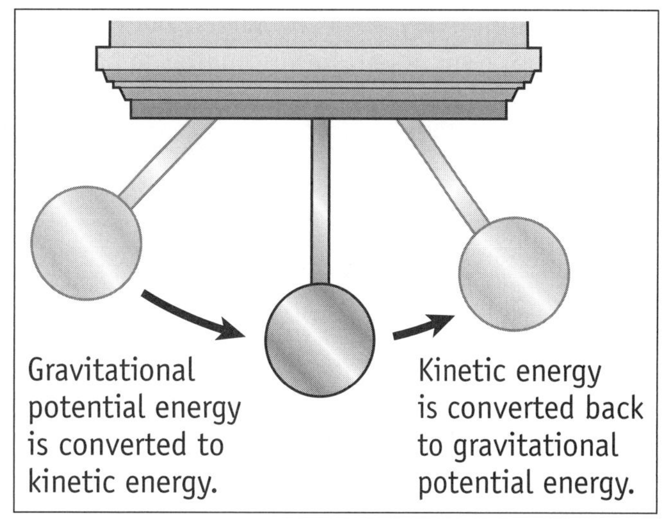

Each swing of the pendulum is slightly shorter than the last swing.

Where did its mechanical energy go? This energy has not disappeared. Part of the pendulum's energy compressed the surrounding air, moving molecules and creating waves; part of the pendulum's energy was changed by the friction of its parts into thermal energy (*heat*). If added together, these different amounts of energy will equal the amount of energy in the pendulum when it began its first swing.

Although both energy and matter follow laws of conservation, there is one important exception: they sometimes can be converted into each other. In nuclear reactions, scientists have found that small amounts of matter can be converted into large amounts of energy.

Because of the possibility of transforming matter into energy or energy into matter, scientists now refer to the law of **conservation of mass — energy**.

WAVES OF ENERGY

Some forms of energy spread in special patterns known as waves. A **wave** is a vibration or disturbance that carries energy through matter or space. Waves generally move away from their source as surface ripples, concentric circles, or spheres. Seismic waves, water waves, **sound waves** and light waves all transfer energy in this form. There are two main types of waves: **mechanical** and **electromagnetic**.

MECHANICAL WAVES

Mechanical waves — such as seismic waves, water waves, and sound waves — always move through some form of matter. Particles in the matter move or vibrate and then pass this energy on to neighboring particles. The type of matter these waves travel through, known as the **medium**, will often affect the properties of the wave. For example, sound waves generally travel faster in solids and liquids than in the air. This occurs because sound waves are caused by the vibrations of particles. Since particles are closer together in solids and liquids than in air, sound vibrations are able to travel faster through them.

ELECTROMAGNETIC WAVES

Electromagnetic waves, such as radio waves or light, do not require a medium. They can travel through many forms of matter, but they can also travel through a vacuum (*open space*). They do not require particles of matter to carry their energy.

DESCRIBING WAVES

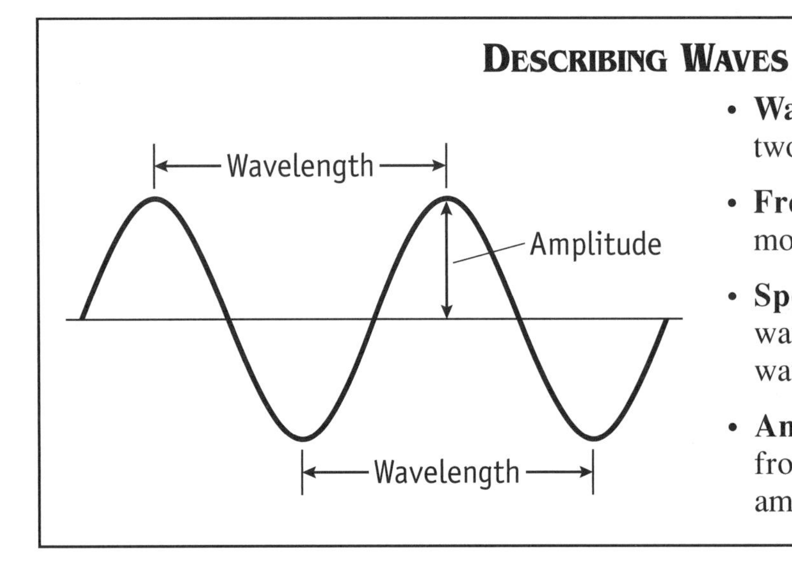

- **Wave length.** The distance between two similar parts of a wave.
- **Frequency.** How many waves move in a given period of time.
- **Speed of Wave.** The speed of a wave equals its frequency times its wavelength.
- **Amplitude.** The height of a wave from its midpoint, showing its amount of energy.

Waves will often change when they move from one medium to another, or when they approach a boundary they cannot pass:

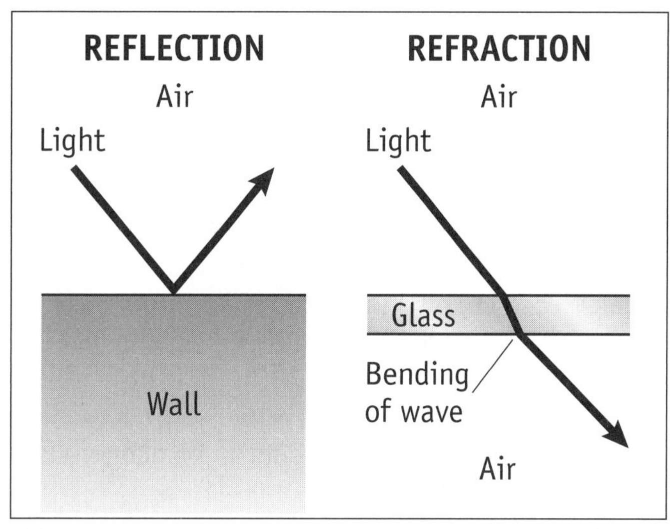

When a wave meets a surface, it can be refracted or reflected.

★ **Reflection.** When the difference between two substances is very great, the wave may not continue into the new medium. Instead, it may bounce back or **reflect**. It continues its same movement but back in the opposite direction. For example, sound waves passing through air will reflect when they suddenly hit a smooth, dense wall. This creates echoes. To reduce echoes, architects and sound engineers introduce softer materials like curtains and carpets. These irregular surfaces absorb the sound waves or reflect them in many different directions, making less echoes.

★ **Refraction.** When waves pass from one medium to another, their speed often changes. This change results in **refraction** — the bending of the waves.

★ **Diffraction. Diffraction** is the bending of waves as they pass through an opening or around a barrier in their path. For example, you can hear someone talking in another room if the door is left open. The sound waves pass through the open doorway and then spread out through diffraction.

★ **Interference.** When waves from two or more sources travel through the same medium at the same time, they strengthen each other if they precisely overlap. Otherwise, they weaken each other and may even cancel each other out completely.

When the source of waves is moving, this can also affect the pattern of the waves, squeezing them in front of the moving source and stretching them out behind. Known as the **Doppler Effect**, this explains why a siren will sound higher in pitch as it approaches you, but lower in pitch after it passes you by. A similar effect with light, known as **red shift**, explains how scientists judge the movement of stars.

THE ELECTROMAGNETIC SPECTRUM

Light waves do not require a medium and can travel through a vacuum, including outer space. They are a part of the **electromagnetic spectrum**, which is made up of visible light waves and other forms of electromagnetic radiation we cannot see. Scientists sometimes think of light and other electromagnetic radiation as streams of little packets of energy known as **photons**. All electromagnetic radiation travels at the same speed, but their frequencies and wavelengths differ. The slowest frequencies on the electromagnetic spectrum have the longest wavelengths; as electromagnetic waves increase in frequency, their wavelengths become shorter.

As you can see from the electromagnetic spectrum below, radio waves have the lowest frequencies. Next come microwaves, used in microwave ovens to cook food. Infrared waves have faster frequencies and carry heat. The heat you feel from a campfire comes from infrared radiation.

Visible light is in the middle of the electromagnetic spectrum. Although we see light as white, Sir Isaac Newton demonstrated that white light is actually made up of several types of waves with different wavelengths. We experience these various wavelengths as different colors. If you shine white light through a glass prism, refraction will separate it into its various colors, because each color bends slightly differently based its wavelength. Violet light has the highest frequency and shortest wavelength of visible light. **Ultraviolet light** has higher frequencies than visible light, with shorter wavelengths. This ultraviolet light can cause sunburn, even on cloudy days. **X-rays** and **gamma rays** have the shortest wavelengths on the spectrum. These high energy waves are used in medicine, but can be dangerous because of their high energies.

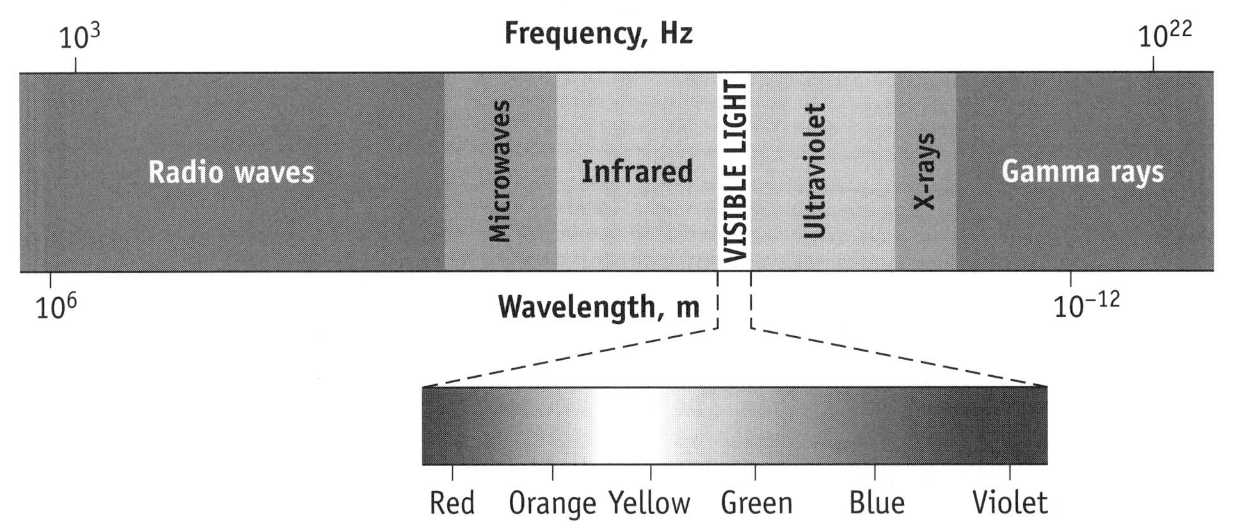

WHAT YOU SHOULD KNOW

- ★ You should know that the forms of energy include **kinetic energy**, the energy of motion, which depends on an object's mass and velocity. **Gravitational potential energy** is the energy stored in an object based on its position on Earth. **Thermal energy** is a form of kinetic energy based on the motion of an object's atoms and molecules and is measured by temperature.
- ★ You should know that energy can be transferred. Thermal energy, for example, can move by conduction, convection, or radiation.
- ★ You should know that energy can change its form: in chemical reactions, chemical energy may be turned into thermal energy; thermal energy can be turned by a machine into mechanical energy. The total amount of energy is conserved.
- ★ You should know that energy is sometimes transmitted in waves. Some **waves** (**seismic, water, sound**) require a medium. **Electromagnetic radiation** can travel through matter or across a vacuum.

CHAPTER STUDY CARDS

Mechanical Energy

★ **Energy.** The ability to do work.

★ **Kinetic Energy.** Energy of motion, $KE = 1/2\ mv^2$

★ **Potential Energy.** Stored energy.

★ **Gravitational Potential Energy.** Potential energy based on an object's position on Earth.

- Gravitational P.E. = weight x height (mgh)
- **Transformation of energy.** Kinetic energy can turn into gravitational energy and back again.

★ **Law of Conservation.** Energy can change its form but its total quantity is always conserved.

Thermal Energy

★ **Thermal Energy.** A form of kinetic energy, based on the vibrations and movements of atoms and molecules.

★ **Temperature.** Measure of the average kinetic energy of an object's particles; the higher the temperature, the greater their movement.

★ **Heat.** Transfer of energy from an object with a higher temperature to one with a lower temperature. Can be transferred by conduction, convention, or radiation.

- **Conduction.** Transfer by direct contact.
- **Convection.** Circular flow of heated fluids.
- **Radiation.** Transfer by waves of energy.

Nuclear Energy

★ **Nuclear Fission.** Splitting of nuclei of large atoms, releasing nuclear energy.

★ **Nuclear Fusion.** Joining together of nuclei of small atoms — hydrogen into helium — also releasing nuclear energy.

★ **Chain Reaction.** When a nucleus is split, its extra neutrons split other nuclei, releasing more energy and nuclei, splitting even more nuclei.

★ **Nuclear Reactions / Conservation of Mass and Energy.** In nuclear reactions, small amounts of matter are converted into large amounts of energy.

Waves

★ **Mechanical Waves.** Seismic, water, or sound — pass through a medium; particles of medium pass along energy of the wave.

★ **Electromagnetic Radiation.** Can pass through some forms of matter but do not require it; can pass through a vacuum or outer space; includes invisible waves as well as visible light

★ **Patterns of Waves.**

- **Refraction.** Wave bends when passing through a different medium.
- **Reflection.** Wave bounces back.
- **Diffraction.** Wave bends around objects.

CHECKING YOUR UNDERSTANDING

1 A cement block is dropped from a helicopter. The block falls 100 feet. Which statement accurately describes the falling cement block?

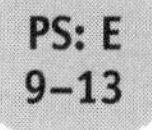

A. Its potential energy decreases as its kinetic energy increases.

B. Its potential energy increases as its kinetic energy decreases.

C. Its potential energy is unchanged as its kinetic energy decreases.

D. Its potential energy is unchanged as its kinetic energy increases.

This question tests your understanding of gravitational potential energy and kinetic energy. Recall what you know about both forms of energy, then apply this knowledge to the question. As the block falls, its speed increases but its height decreases. Therefore, its kinetic energy increases while its potential energy decreases. Thus, the correct answer can only be choice A.

Now try answering some questions on your own about the nature of energy.

2 The kinetic theory states that the higher the temperature, the faster the

PS: E 9–11

A. lighter particles within a substance clump together
B. bonds within atoms break down
C. particles that make up a substance move
D. molecules of gas rush together

- Examine the Question
- Recall What You Know
- Apply What You Know

3 The diagram shows a compressed spring between two carts at rest on a frictionless surface. When the spring is released, the two carts move in opposite directions. Cart A has a mass of 2 kilograms, and cart B has a mass of 1 kilogram. What will occur when the spring is released?

PS: E 9–12

A. The kinetic energy of cart A will equal that of cart B.
B. The kinetic energy of cart A will equal the potential energy of cart B.
C. The kinetic energy of cart A will increase as the potential energy of the spring decreases.
D. The kinetic energy of cart A will be less than the potential energy of cart B.

4 Which type of radiation in the electromagnetic spectrum mainly transfers thermal energy?

PS: G 9–18

A. x-rays
B. infrared waves
C. ultraviolet waves
D. radio waves

5 Which graph represents the relationship between the gravitational potential energy of an object and the object's height above the surface of Earth?

PS: E 9–13

Gravitational potential energy
Height
A.

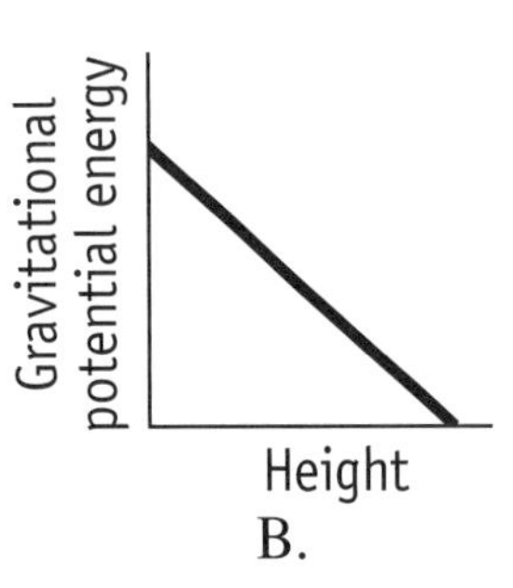

B.

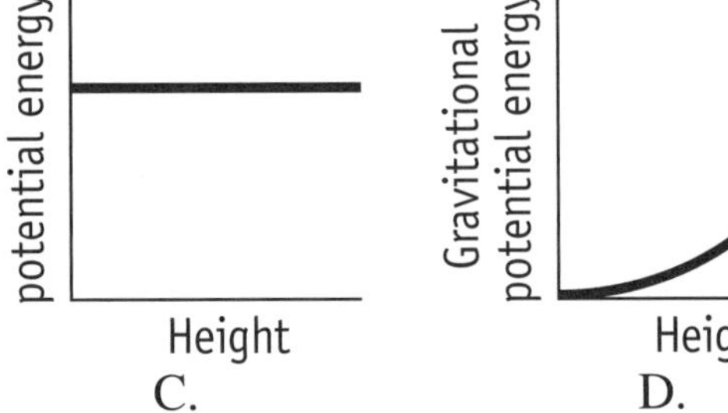
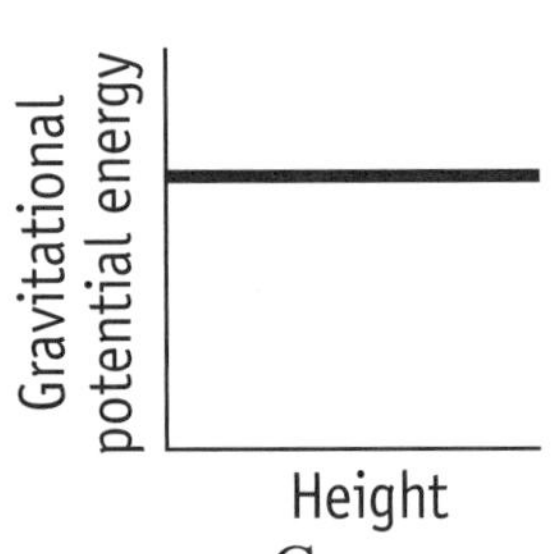

C.

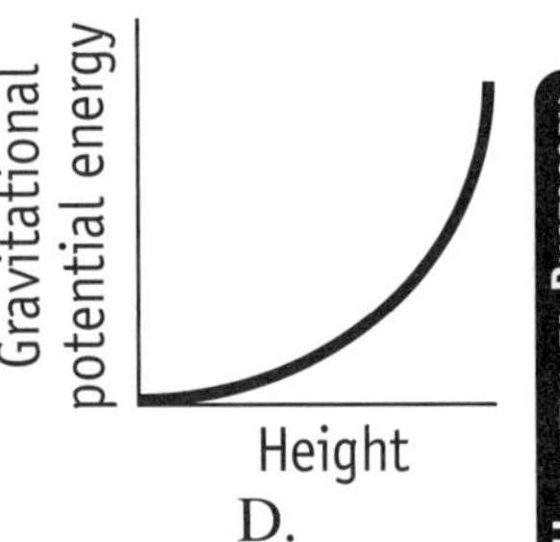

D.

6 **A student prepared a large box in which several dozen ping pong balls were placed on individually set mousetraps. The student then dropped one ping pong ball into the box. This triggered one of the mousetraps, which in turn triggered other mousetraps, triggering even more mousetraps. Which of the following has the student demonstrated?**

A. alpha decay
B. fusion
C. chain reaction
D. theory of relativity

PS: F 9–14

7 **Lemon juice, a weak acid, is added to baking soda. Heat from the surroundings is absorbed by the reactants during the reaction. The reaction is therefore**

A. interrupted
B. ionized
C. exothermic
D. endothermic

PS: F 9–16

8 **A student shines a ray of light through a triangular glass prism. The white light separates into different colors. Which process could best be used to explain their separation?**

A. reflection
B. refraction
C. diffusion
D. diffraction

PS: G 9–20

9 **A team of research scientists bombard the nuclei of Uranium-235 atoms with neutrons. The nuclei of the uranium atoms divide, emitting free neutrons and energy. This process is known as**

A. fusion
B. fission
C. covalent bonding
D. convection

PS: F 9–14

10 **Scientists define energy as the ability to perform work. There are many forms of energy. List two types of energy and provide an example of each type. (*4 points*).**

- ♦ Examine the Question
- ♦ Recall What You Know
- ♦ Apply What You Know

PS: E 9–15

11 **A student adds ice cubes to a beaker containing cold water. What happens to the water molecules in the water?**

A. Their bonds weaken and they separate into oxygen and hydrogen atoms.
B. They increase in speed because of convection currents.
C. Their heat melts the ice by radiation.
D. They decrease in speed as they transfer energy to the ice by conduction.

PS: F 9–17

12 **Scientists are preparing to launch a satellite into outer space. Which form of energy will the scientists be able to use to communicate with that satellite?**

A. sound waves
B. sonar
C. radio waves
D. electricity

PS: G 9–18

13 Scientific theories and discoveries are often able to change how people view the world. Select two of the following examples, and explain how science has brought about a change in people's views:

- Newton's laws of motion
- The methods used by people to cook food
- The use of a prism to separate visible light
- The conversion of matter into energy in nuclear reactions (*2 points*)

◆ Examine the Question
◆ Recall What You Know
◆ Apply What You Know

PS: H 9–26

14 Keesha mows her neighbor's lawn with a gasoline-powered lawn mower. Which describes some of the transformation of energy that occurs in the mower?

PS: F 9–15

A. chemical energy ➜ kinetic energy ➜ thermal energy
B. gravitational potential energy ➜ kinetic energy ➜ chemical energy
C. electrical energy ➜ chemical energy ➜ kinetic energy
D. thermal energy ➜ kinetic energy ➜ nuclear energy

15 An architect is designing a concert hall. Which design can the architect adopt to reduce echoes in the new concert hall?

PS: G 9–20

A. use of curtains, carpets, and textile prints on the walls
B. use of smooth metal surfaces to cover the walls and ceiling
C. use of a round tiled dome for the ceiling
D. use of polished mirrors on the walls and hardwood floors

Use the chart below to answer the following question

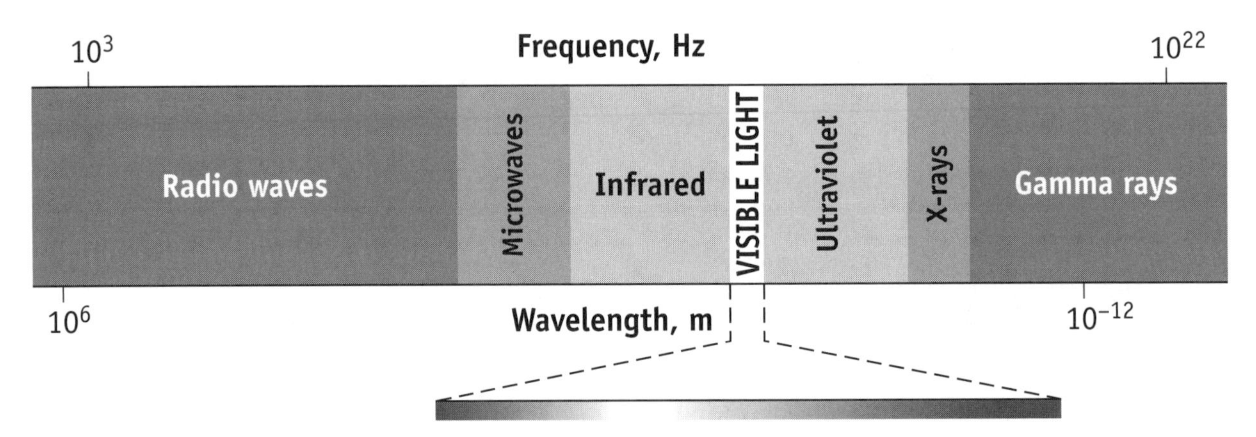

16 One could conclude from this chart that the wavelengths of electromagnetic waves

PS: G 9–18

A. decrease as their frequency increases
B. increase as their frequency increases
C. remain the same as their frequency increases
D. are unrelated to frequency

CHECKLIST OF BENCHMARKS IN THE PHYSICAL SCIENCE UNIT

Directions. Now that you have completed this unit, place a check mark (✔) next to those benchmarks you understand. If you have trouble recalling information connected with one of the benchmarks, review the chapter indicated in the brackets for the items you do not recall.

- ❑ You should be able to explain how matter is made of minute particles called atoms, and how atoms are comprised of even smaller components. You should also be able to explain the structure and properties of atoms. **[Chapter 11]**

- ❑ You should be able to explain how atoms react with each other to form substances, and how ions and molecules react with each other or other atoms to form even different substances. **[Chapter 12]**

- ❑ You should be able to describe the identifiable physical properties of substances (*e.g., color, hardness, conductivity, density, concentration and ductility*). You should also be able to explain how changes in these properties can occur without changing the chemical nature of the substance. **[Chapter 12]**

- ❑ You should be able to explain the movement of objects by applying Newton's three laws of motion. **[Chapter 13]**

- ❑ You should be able to demonstrate that energy can be considered to be either kinetic (*motion*) or potential (*stored*). **[Chapter 14]**

- ❑ You should be able to explain how energy can change its form or be redistributed, but that the total quantity of energy is conserved. **[Chapter 14]**

- ❑ You should be able to demonstrate that waves (*e.g., sound, seismic, water and light*) have energy, and that waves can transfer energy when they interact with matter. **[Chapter 14]**

- ❑ You should be able to trace the historical development of scientific theories and ideas, and describe emerging issues in the study of physical sciences. **[Chapters 11, 12, 13, and 14]**

LIFE SCIENCES

In this unit, you will review what you need to know for the **OGT in Science** about the Life Sciences — those sciences that study living things. You will learn how the cell is the basic unit of living things. You will also explore heredity, the role of the DNA molecule in life, the processes of evolution, and how living things interact with their environment.

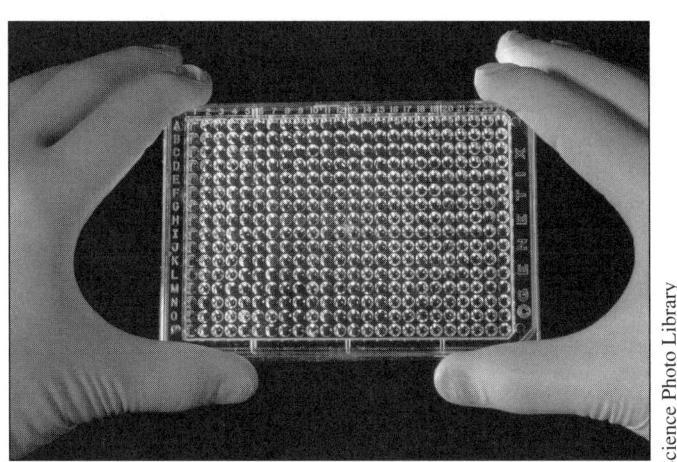

A tray containing human genome DNA

Science Photo Library

★ **Chapter 15: Cells and Cellular Processes**

In this chapter, you will learn about cell theory. This chapter will show you how to identify different types of cells and their parts. In addition, you will study cellular processes, how cells use energy and matter, and how cells grow and divide.

★ **Chapter 16: Heredity and Genetics**

This chapter explains how cells inherit many characteristics. You will learn about genes, chromosomes, and DNA — the molecules that carry the instructions that determine who we are. You will also review the work of important scientists in the field of genetics — Gregor Mendel, Francis Crick, and James Watson.

★ **Chapter 17: Evolution**

This chapter focuses on the mechanisms and processes of biological evolution and how evolution has contributed to the wide variety and diversity that exists among life forms.

★ **Chapter 18: Ecology**

In this chapter, you will learn about various types of ecosystems, how ecosystems change over time, how energy and matter flow through an ecosystem, and how human activities affect existing ecosystems.

CHAPTER 15

CELLS AND CELLULAR PROCESSES

MAJOR IDEAS

A. The cell is the basic unit of all living things. All existing cells come from pre-existing cells.

B. The oldest form of cells, **prokaryotic cells**, have DNA but no nucleus. They divide through binary fission.

C. **Eukaryotic cells** have a membrane-covered nucleus. They are found in unicellular and multicellular organisms. They divide through mitosis and meiosis.

D. Plant cells have cell walls, vacuoles, and chloroplasts for photosynthesis. Plant cells convert the energy of sunlight into chemical energy through **photosynthesis**. Cellular respiration, taking place in a cell's mitochondria, releases this energy for cellular purposes. **ATP** acts as the "currency of energy" for cells.

WHAT IS A CELL?

The discovery of **cells** only came about after the invention of the microscope. In 1665, the English scientist Robert Hooke identified the first cells when he looked at a piece of cork through a microscope. Since Hooke first peered through his microscope, scientists have found that a cell is the smallest unit of matter capable of carrying on all the processes of life. We now know that all living cells share certain common characteristics:

★ They are surrounded by a **cell membrane**.

★ They contain hereditary material in the form of **DNA** (*a large organic molecule*), which they receive from a pre-existing cell or cells.

★ They are mainly composed of fluid, which contains chemicals and structures that allow them to live, grow and reproduce. The fluid and its structures are known as **cytoplasm**.

★ They are composed of a small number of key chemical elements — *carbon*, *hydrogen*, *oxygen*, *nitrogen*, *phosphorus*, and *sulfur*.

Modern **cell theory** consists of three main parts:

★ The cell is the basic unit of structure and function of all living organisms.

★ All living things are made up of cells, which can be of various types.

★ All living cells come from pre-existing cells.

Viruses, discovered in 1935, do not have the same characteristics as cells. They cannot grow, or carry on cellular functions (*like homeostasis*). They can only reproduce in a host cell. Although they have RNA or DNA, like a cell, scientists do not consider them to be living things. They are chemicals that enter living cells and multiply.

PROKARYOTIC AND EUKARYOTIC CELLS

There are two main types of cells: **prokaryotes** and **eukaryotes**.

PROKARYOTIC CELLS

The oldest cells on Earth are **prokaryotic** (proh-KAR-ee-OHT-IK) **cells**. These cells are very small. What distinguishes a prokaryotic cell is that it has no cell nucleus and none of the structures inside the cell have membranes.

All prokaryotic cells are bacteria. Because they have no membrane-covered nucleus, their DNA is just bunched up like a curled rubber band in the cytoplasm. It forms one long, continuous, circular chromosome. Most prokaryotic cells also have a cell wall outside the cell membrane, and ribosomes in the cytoplasm. These are unicellular (*one-celled*) organisms.

PROKARYOTIC CELL STRUCTURE

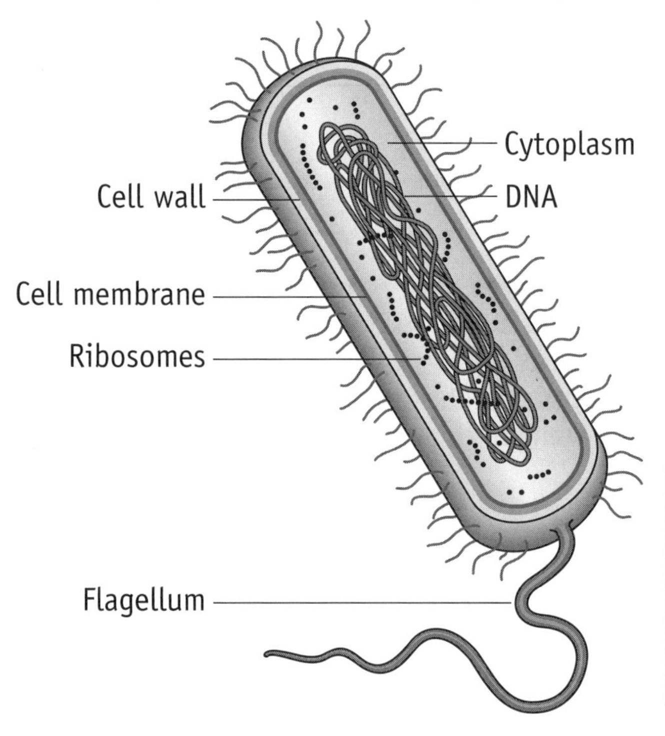

EUKARYOTIC CELLS

All other cells are known as **eukaryotic** (yoo-KAR-ee-OHT-IK) cells. These cells are generally ten times the size of prokaryotic cells or larger. The defining characteristic of eukaryotic cells is that each has a membrane-covered nucleus.

Many eukaryotic cells are unicellar — such as amoeba and paramecia. However, other eukaryotic cells form parts of multicellular organisms. These organisms have many cells, with specialized functions.

In a eukaryotic cell, the cell's DNA is stored in the nucleus. A eukaryotic cell also has other specialized structures in its cytoplasm, known as **organelles**. These organelles allow the cell to conduct several different chemical processes at the same time. Eukaryotic cells have the following types of structures:

EUKARYOTIC CELL

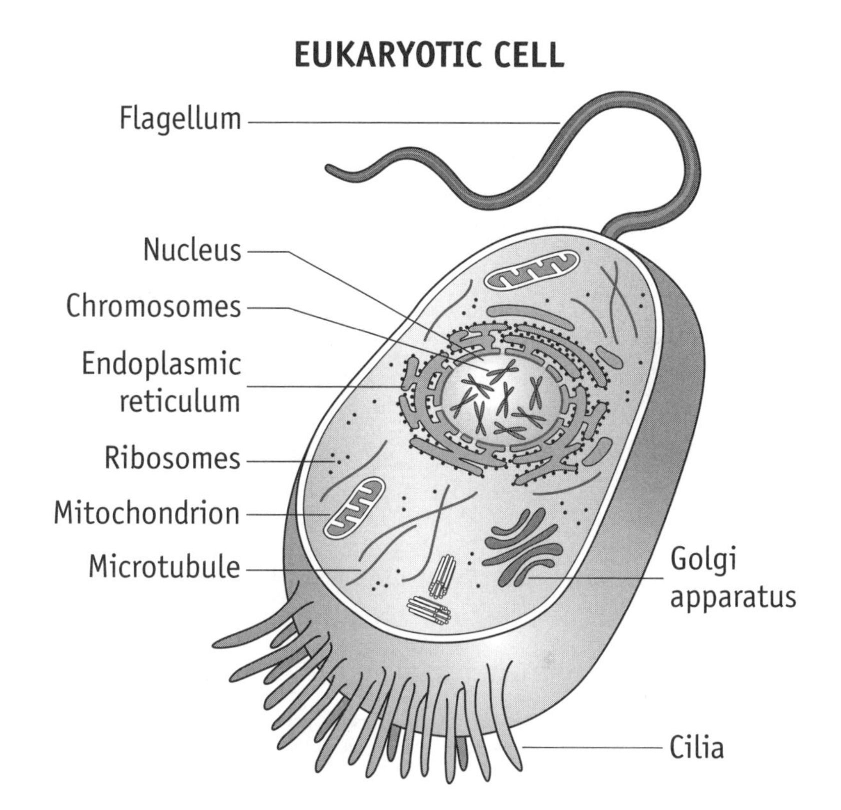

- ★ **Cell Membranes** hold the cell together and determine what enters and leaves the cell. The cell membrane is made up of two layers of lipid molecules (*phospholipids*), with embedded proteins.
- ★ Some cells are propelled by tiny hairs known as **cilia**, or a cellular "tail," known as a **flagellum**. Cilia and flagella are both made of hollow tubes known as **microtubules**.
- ★ **Cytosol** is the fluid in the cytoplasm.
- ★ The **nucleus** is the largest organelle in a eukaryotic cell. It is surrounded by a double membrane. In order to live and grow, cells must synthesize proteins, lipids, carbohydrates and other organic compounds. The cell's nucleus stores the cell's DNA, and makes RNA and ribosomes. The DNA in the cell's nucleus directs cell functions.
- ★ **Ribosomes** are made up of proteins and RNA. They provide a place for the cell's synthesis of proteins, which are used by the cell. Unlike other organelles, ribosomes are not covered by any membrane.
- ★ The **endoplasmic reticulum** consists of membranes and tubes that act as an intercellular path to transport proteins and other molecules inside the cell or outside it.

★ The **Golgi complex** consists of membranes that process, package and secrete proteins and other molecules from the cell.

★ **Mitochondria** are in charge of cellular respiration. They are surrounded by a double membrane. The inner membrane has many folds. Within these folds, chemical reactions take place transferring energy from organic compounds to ATP (*adenosine triphosphate*) — the molecule cells use to store energy.

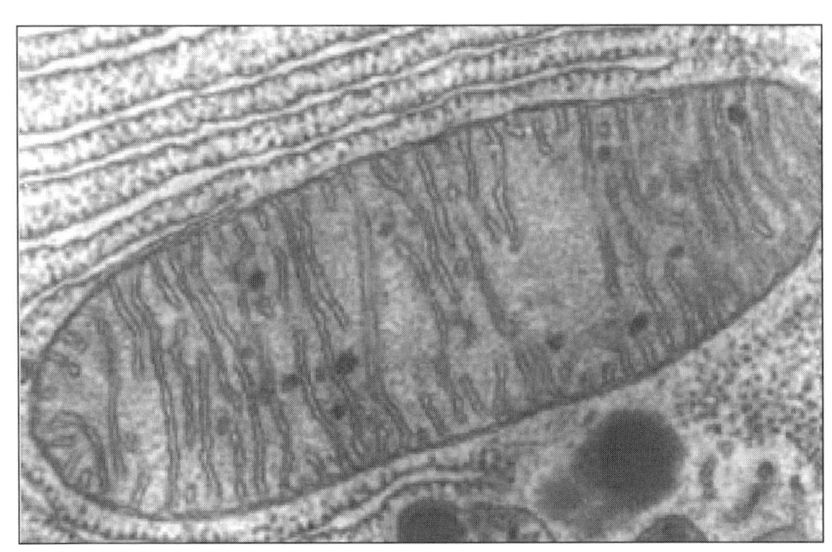

A mitochondria as seen under an electron microscope.

★ **Lysosomes** are organelles found in animal cells and a few plant cells. They contain enzymes that dissolve food, foreign invaders, or waste.

Besides these structures, plant cells also have the following special features:

★ **Cell Walls.** The cell membranes of plants are covered by a rigid, outer cell wall made of cellulose that helps support and protect the plant.

★ **Vacuoles.** These organelles are large fluid-filled "bubbles" within plant cells that hold water, enzymes and waste. Often the vacuole takes up most of the space in a plant cell.

★ **Plastids.** Similar to mitochondria, plastids have two membranes. The most common type of plastid, the **chloroplast**, is responsible for *photosynthesis* — where sunlight is converted into chemical energy.

HOW ANIMAL CELLS DIFFER FROM PLANT CELLS

Animal Cells	Plant Cells
• Can move about and divide	• Are confined by cell walls
• Must consume other organisms for food	• Have chloroplasts to conduct photosynthesis
• Have lysosomes	• Have large vacuoles for waste

APPLYING WHAT YOU HAVE LEARNED

✦ Make your own drawing of a typical eukaryotic cell, and label its main parts.

CELLULAR PROCESSES

In order to live, grow and reproduce, cells must carry out certain cellular processes.

HOMEOSTASIS

To survive, cells and all living things must maintain stable internal conditions. For example, cells must control how much water they have, their temperature, and their pH level. The cell must balance what enters and leaves it, and control the pace of its chemical reactions. **Homeostasis** is the process of keeping this **equilibrium** (*dynamic balance*).

TRANSPORTATION OF MOLECULES

A cell controls what enters and leaves it through passive and active transport.

★ **Passive Transport** requires no energy from the cell. It results from random molecule motion. Many molecules enter or leave the cell through **diffusion** — the simple movement of molecules across the cell membrane from areas of high concentration to areas of low concentration. The movement of water molecules from an area of high concentration to one of lower concentration is known as **osmosis**. Certain "carrier proteins" in the cell membrane can aid in diffusion by binding to molecules.

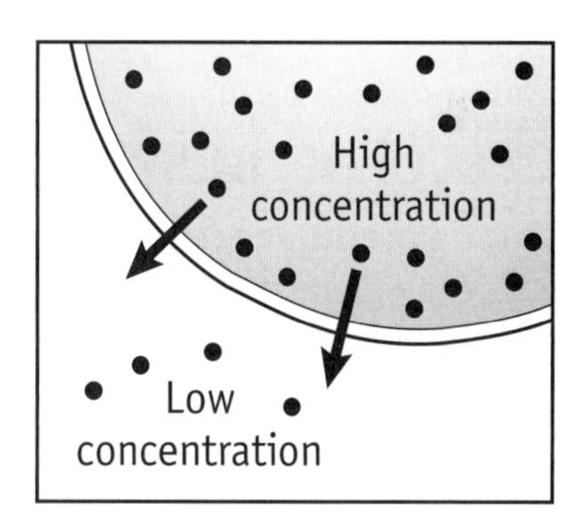

★ **Active Transport**. Sometimes carrier proteins help move small particles across cell membranes in the opposite direction from diffusion. Energy from the cell is required in this case, which it obtains from ATP. A cell can also ingest fluid or larger particles in the processes known as **endocytosis**. The cell membrane engulfs the particle it is trying to ingest. Once the particle is completely enclosed in the membrane, it is pinched off in a membrane-covered sac and enters the cell.

THE PROCESS OF ENDOCYTOSIS

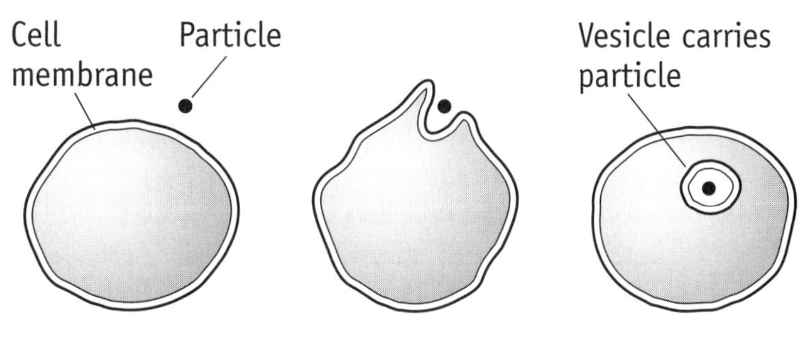

The opposite occurs in **exocytosis**. The endoplasmic reticulum or Golgi complex wraps the particle in membrane and moves it outside the cell.

HOW CELLS GAIN AND RELEASE ENERGY

All forms of life require energy to carry out the functions of life. Plants acquire energy from the sun through photosynthesis. Other organisms acquire energy by eating plants or by eating other organisms that eat plants — obtaining organic compounds with stored energy.

PHOTOSYNTHESIS

Plants obtain energy from sunlight and store it for later use in the bonds of organic compounds. This process of transferring energy is known as **photosynthesis**. The chloroplasts in plants have light-sensitive green pigment called **chlorophyll**. Chlorophyll aborbs light, which affects its electrons. Through a series of complex chemical reactions, plants use the energy from sunlight to change carbon dioxide (CO_2) and water (H_2O) into glucose ($C_6H_{12}O_6$) and oxygen (O_2).

$$6CO_2 + 6H_2O + \text{light energy} \rightarrow C_6H_{12}O_6 + 6O_2$$

CELLULAR RESPIRATION

Through **cellular respiration**, cells break down nutrients to release energy. Cellular respiration is not the same as breathing, but there are similarities: both require oxygen and produce carbon dioxide. Cells use oxygen in a complex series of reactions to convert glucose into CO_2 and H_2O. These reactions release the energy stored in the glucose by photosynthesis. In fact, cellular respiration is almost the reverse of photosynthesis.

$$C_6H_{12}\,O_{12} + 6O_2 \rightarrow 6CO_2 + 6H_2O + \text{energy}$$

This released energy is stored by the cell in the form of **ATP**. ATP acts as the cell's "energy currency." The ATP molecule has three phosphate groups. The last phosphate group can break off, releasing energy that can be transferred to another chemical reaction at the same time. For this reason, cells use ATP molecules to store and release energy.

$$\text{ATP} \rightleftarrows \text{ADP} + \text{P} + \text{energy}$$

FERMENTATION

Fermentation occurs when cells make ATP without oxygen by changing glucose into a different organic compound. Because no oxygen is used, fermentation is **anaerobic** and produces much less energy than cellular respiration. Fermentation occurs when yeast cells, for example, turn sugar into alcohol and carbon dioxide. Another example of fermentation occurs when our muscle cells cannot get enough oxygen and use fermentation to get energy.

APPLYING WHAT YOU HAVE LEARNED

✦ Make a chart comparing ***photosynthesis***, ***cellular respiration***, and ***fermentation***.

CELLULAR DIVISION

All living cells come from pre-existing cells. All cells undergo a recurring pattern of development known as the **cell cycle**. Cells reproduce through the process of cell division. Cell division requires **replication** (*copying*) of the cell's **DNA** — the hereditary instructions that tell the cell how to function.

PROKARYOTES

Prokaryotic cells divide through the process of **binary fission**. The cell's DNA attaches to the cell membrane and copies itself. Then the cell grows to twice its size. A new cell wall develops, separating the two new cells. Each new cell is identical to the original cell.

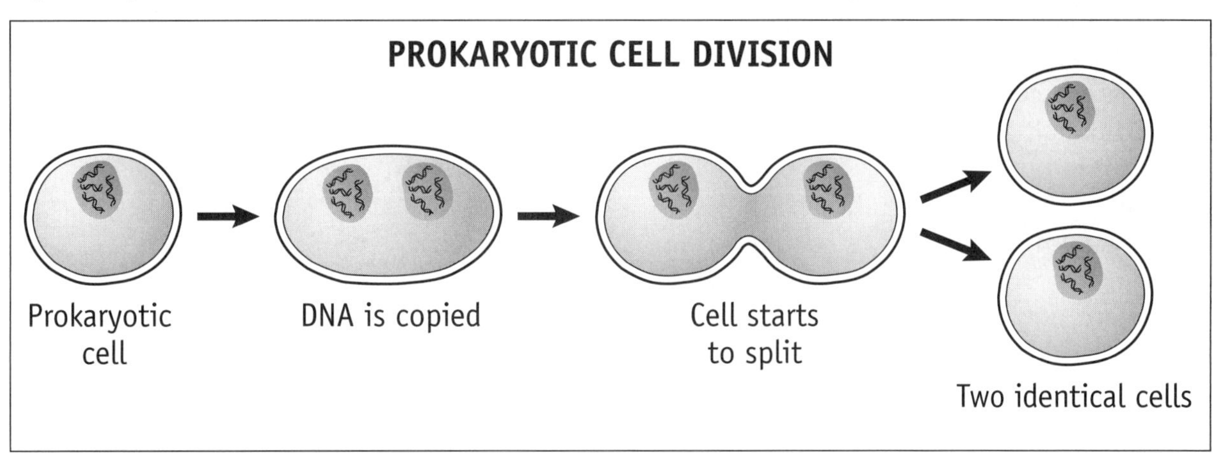

EUKARYOTES: MITOSIS AND CYTOKINESIS

The division of eukaryotic cells is more complex. The cell's nucleus divides in the process known as **mitosis**. The DNA is densely packed in rod-shaped **chromosomes**. For example, human cells each have 46 chromosomes, which contain six billion pairs of organic compounds known as **nucleotides**. During mitosis, the chromosomes copy themselves but stay attached at the center. Each chromosomal copy is known as a **chromatid**. Then, the nuclear membrane disappears, and **centrosomes** (*dark spots*) appear.

The centrosomes migrate to opposite ends of the cell. **Spindle fibers**, made from microtubules, attach from the centrosomes to the middle of the cell. Next, chromosomes line up at the end of the spindle fibers across the middle of the cell. The chromatids separate, pulled by the spindle fibers towards each of the centrosomes. When they reach the centrosomes, the spindle fibers fall apart and a membrane forms around each new nucleus.

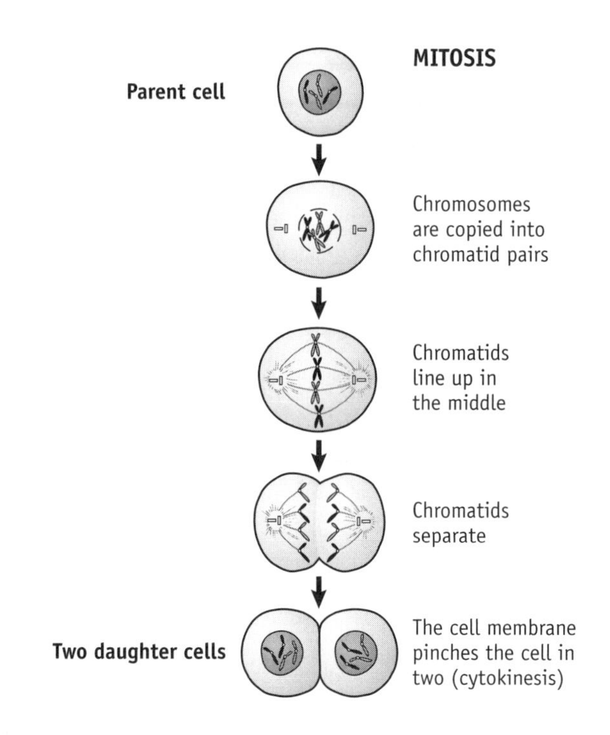

With this division, each new cell has an identical copy of the original cell's chromosomes. At the same time, the cell's cytoplasm divides in the process known as **cytokinesis**. In animal cells, the cell membrane pinches and separates the cell in two. In plant cells, a new cell wall develops, separating the divided cells.

MEIOSIS

The sex cells of animals and plants divide through **meiosis**, a special process that allows for genetic variation. During meiosis, a cell's chromosomes are reduced in half.

This occurs because the new cells later join with other sex cells during *fertilization* to produce offspring with some characteristics from each parent. For example, both human sperm and egg cells are produced through meiosis. If the number of chromosomes were not reduced during meiosis, when an egg and sperm join the new cell would have too many chromosomes.

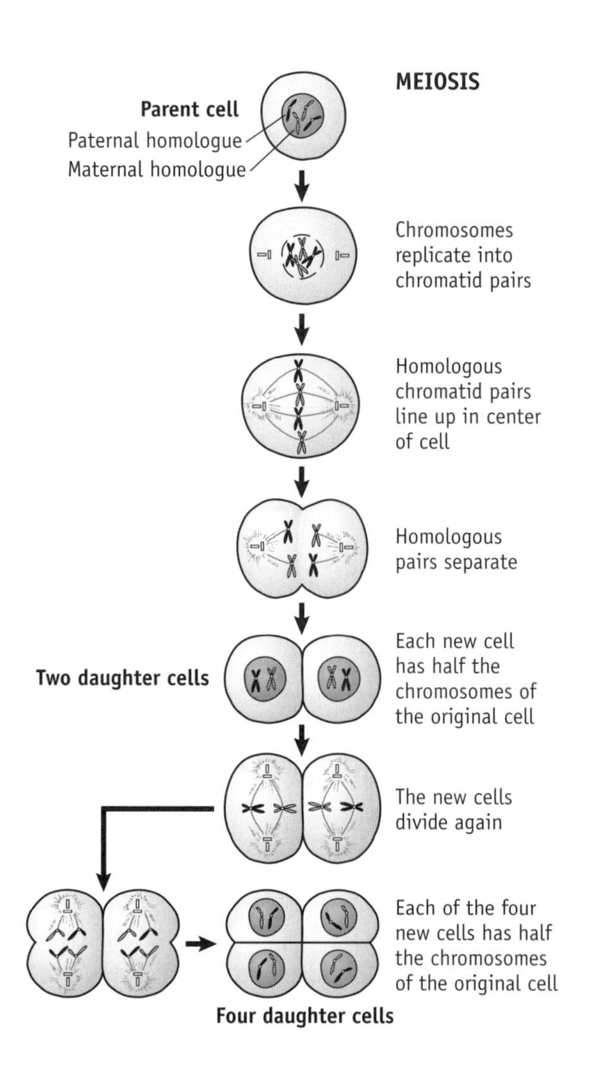

THE PHASES OF MEIOSIS

Homologous Chromosomes. In humans and other organisms, chromosomes come in pairs. Each chromosome pair is made up of two chromosomes of similar size and shape, governing the same traits, such as hair color. These pairs are referred to as **homologous chromosomes**. These homologous chromosomes play an important role in meiosis.

Chromatid Pairs. Before meiosis begins, chromosomes duplicate themselves into chromatid pairs. In the first phase of meiosis, spindle fibers appear, the nuclear membrane dissolves, and each chromatid pair lines up with its homologous chromatid pair. The two chromatid pairs twist around, swapping some genes. This process contributes to genetic variation.

The homologous pairs then line up randomly along the middle of the cell. Next they are pulled to opposite ends of the cell. Each new cell is left with half the number of ordinary chromosomes. It receives only one chromosome from each homologous chromosome pair. However, each new cell has two chromatid copies for each of its chromosomes.

Final Phase of Meiosis. The two new cells now divide again, making four cells. As each cell divides, its chromatids separate so the number of chromosomes each new cell gets remains the same. Each of the four new cells therefore ends up with half the number of chromosomes of the original parent cell. A nuclear membrane forms around the chromosomes in each of the four new cells. Later, these special cells will be joined with other sex cells formed by meiosis to produce new offspring. A human egg and sperm, for example, each has 23 chromosomes. When they join together, they produce a cell with 46 chromosomes. In this way, genes from both parents are inherited. You will learn more about heredity in the next chapter.

APPLYING WHAT YOU HAVE LEARNED

- ✦ Using a chart, compare mitosis and meiosis. Show:
  - How are they different?
  - How are they similar?

WHAT YOU SHOULD KNOW

- ★ Make sure you know the basic idea of cell theory — that all living things are made of cells, and that all cells come from pre-existing cells.
- ★ Make sure you know the main differences between prokaryotic and eukaryotic cells.
- ★ Make sure you know the special structures unique to plant cells.
- ★ Make sure you know how cells obtain and use energy, and that you are able to explain the processes of photosynthesis, cellular respiration and fermentation.
- ★ Make sure you know how cells divide by either binary fission, mitosis, or meiosis.

CHAPTER STUDY CARDS

Cell Theory

Modern cell theory consists of three parts:

★ **Basic Unit of Living Things.** The cell is the basic unit of all living things.

★ **Living Things Are Made of Cells.** All living things are made up of cells,

★ **Come From Pre-existing Cells.** All living cells come from pre-existing cells.

The Two Main Types of Cells

★ **Prokaryotic Cells.** A bacteria cell with no nucleus and none of the structures inside the cell have membranes. However, these cells have DNA and a cell wall outside the cell membrane.

★ **Eukaryotic Cells.** These include all other cells. They are characterized by a membrane-covered nucleus; their DNA is stored in the cell's nucleus; they have additional organelles including mitochondria for cellular respiration.

Eukaryotic Cell Structures

★ **Cell Membrane** holds cell together.

★ **Cytosol** is the fluid in cytoplasm.

★ **Nucleus** contains DNA, makes RNA.

★ **Ribosomes** is where proteins are made.

★ **Mitochondria** conducts respiration.

★ **Endoplasmic Reticulum** transports molecules.

★ **Golgi Complex** helps secrete molecules.

★ **Lysosomes** contain enzymes.

Cellular Processes

★ **Homeostasis.** a cell must maintain stable internal conditions.

★ **Transportation of Molecules / Waste** A cell controls what enters and leaves it by passive and active transport.

- ***Passive Transport.*** Molecules enter or leave the cell through diffusion, moving from areas of high to low concentration.
- ***Active Transport.*** Helps move small particles across cell membranes in the opposite direction of diffusion; requires use of energy by the cell.

How Cells Acquire and Release Energy

★ **Photosynthesis.** The chemical process in the chloroplasts by which plants use light to convert carbon dioxide and water into carbohydrates, releasing oxygen.

★ **Cellular Respiration.** The process in the mitochondria which breaks down glucose into usable energy; the reverse of photosynthesis.

★ **Fermentation** occurs when cells make ATP without oxygen (*anaerobic*) by changing glucose to a different organic compound.

Cell Division

★ **Prokaryotes.** These cells divide by binary fission, in which each new cell is identical to the original cell.

★ **Mitosis and Cytokinesis.** In mitosis, the cells' nucleus divides. Each of the new cells has an identical copy of the original cell's chromosomes.

★ **Meiosis.** A process of sex cell division that allows for genetic variation. Cells are left with half the chromosomes of the original cell. This produces sperm and eggs for sexual reproduction.

CHECKING YOUR UNDERSTANDING

1 **Because of a defect, a cell has lost its ability to regulate the passage of water, food, and wastes into and out of the cell. In which structure of the cell is this defect most likely located?**

LS: B
10–2

A. the ribosomes
B. the cell membrane
C. the cytosol
D. the endoplasmic reticulum

You must identify the cell structure that regulates the transportation of water, food and wastes into and out of a cell. Choice A is incorrect since ribosomes direct the cell's synthesis of proteins. From the structures remaining, which one determines what enters and leaves the cell?

Now answer some questions on your own about cells and cell division

2 **One difference between plant and animal cells is that animal cells lack**

LS: A
10–1

A. a nucleus
B. chloroplasts
C. a cell membrane
D. lysosomes

3 **Which statement is not a part of the cell theory?**

LS: A
10–1

A. Cells are the basic unit of structure of living things.
B. There are a variety of types of cells.
C. Cells can be spontaneously created from compounds.
D. All cells come from preexisting cells.

4 **The ability of cells to pass on their characteristics to new cells is directly related to the ability of**

LS: C
10–5

A. cytoplasm to expel wastes
B. DNA to attach to a cell
C. ribosomes to use energy
D. chromosomes to replicate

5 **Which phrase describes cellular respiration, a process continuously occurring in the mitochondria of cells?**

LS: D
10–10

A. removal of all oxygen from the cells of an organism
B. the breakdown of organic compounds to obtain energy
C. transport of materials within cells and throughout the bodies of multicellular organisms
D. a reaction that stores energy in glucose

- Examine the Question
- Recall What You Know
- Apply What You Know

Use the illustration below to answer the following question.

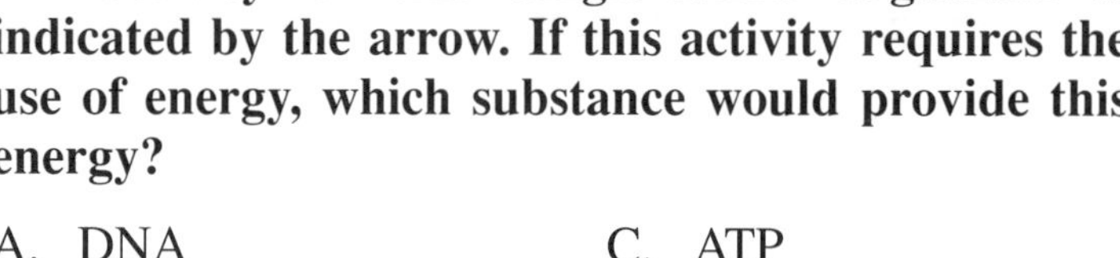

6 An activity of this single-celled organism is indicated by the arrow. If this activity requires the use of energy, which substance would provide this energy?

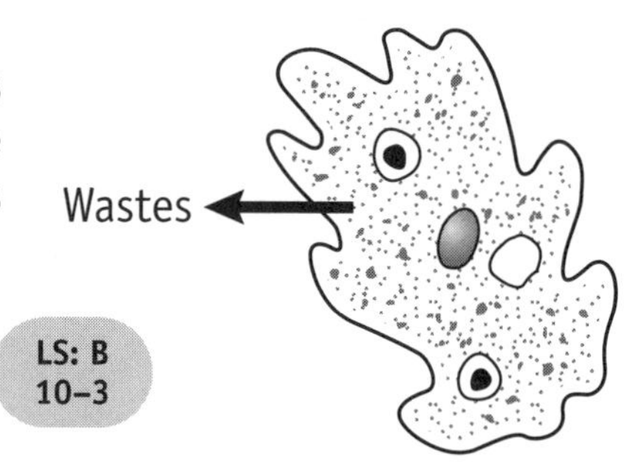

A. DNA
B. a hormone
C. ATP
D. an antibody waste

LS: B 10–3

Use the diagram below to answer the following two questions.

7 This diagram represents a unicelled organism in a watery environment. If the ▲ symbols represent molecules of a specific substance, what process does arrow "A" represent?

A. active transport
B. vacuoles
C. photosynthesis
D. fermentation

LS: B 10–3

8 What does arrow "B" represent?

A. cytosol
B. cell membrane
C. nucleus
D. ribosomes

LS: B 10–2

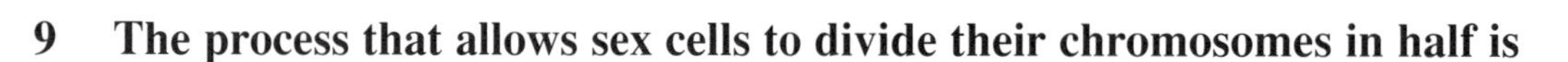

9 The process that allows sex cells to divide their chromosomes in half is

A. fermentation
B. binary fission
C. mitosis
D. meiosis

LS: B 10–4

10 The diagram to the right represents a process that often occurs in the cell of a plant. According to this diagram, what process is being illustrated?

A. fermentation
B. photosynthesis
C. mitosis
D. binary fission

LS: D 10–10

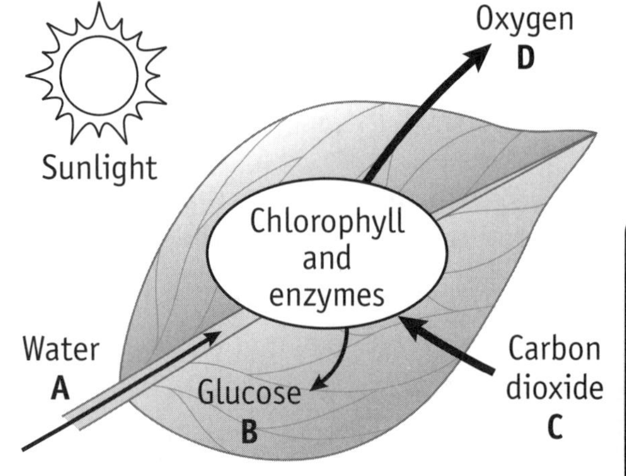

CHAPTER 16

HEREDITY AND GENETICS

In this chapter, you will learn about genetics — the study of how hereditary traits are passed from parents to their offspring, and how DNA provides the basis of inheritance.

MAJOR IDEAS

A. Hereditary traits are determined by genes. Two genes, or alleles, govern each trait. In sexual reproduction, an offspring receives an allele from each parent for each trait.

B. A dominant trait appears in an organism if it is present in either allele. A recessive trait appears only if it is carried by both alleles.

C. Genes determine an organism's sexual identity: females have two X chromosomes (XX); males have an X and a Y chromosome (Xy). Traits are sex-linked if they are carried on the X or Y chromosome.

D. DNA molecules carry genetic information within each chromosome. Their structure makes it easy for them to carry information and to copy themselves. Spontaneous changes in DNA, known as mutations, are a source of genetic variation.

MENDEL AND THE LAWS OF GENETICS

In the 1850s, **Gregor Mendel**, an Austrian monk, became the first scientist to discover the laws of genetics. Mendel conducted his experiments on pea plants. He found certain characteristics of pea plants occurred in pairs of contrasting traits. For example, pea plants may be tall or short; their flowers are purple or white; their seeds are smooth or wrinkled; and their seeds are green or yellow. Mendel was able to **self-pollinate** plants by rubbing pollen on the same plant's flowers. He could **cross-pollinate** plants by rubbing pollen from one plant on the flowers of another plant.

Mendel self-pollinated his plants several times until he was sure they were pure for each trait. Then he cross-pollinated them. He came to some surprising results.

Mendel found that for each pair of contrasting traits, one trait was **dominant**. For example, if he crossed a pea plant having purple flowers with a plant containing white flowers, the offspring always had purple flowers. If he crossed a pea plant with smooth seeds with one with wrinkled seeds, the offspring always had smooth seeds. However, he then found the contrasting trait re-appeared. Scientists refer to this trait as **recessive**. The recessive trait later reappeared in one-fourth of the plants after the first generation. From these results, Mendel concluded that each trait was governed by two factors. One factor was inherited from each parent. The recessive trait only appeared in the plant if it was inherited from *both* parents. If the plant inherited the dominant trait from one of the parents, it "masked" or hid the recessive trait.

Gregor Mendel

Mendel also concluded that each characteristic was determined independently. This is known as the **Law of Independent Assortment**. The type of seed a pea plant has does not affect the color of its flowers.

Scientists now refer to Mendel's "factors" as genes. A **gene** is that part of a chromosome that determines a specific trait — such as hair or eye color, or whether a pea plant has a smooth or wrinkled seed. The two genes that determine the same trait are known as **alleles**. Thus, one pair of alleles governs whether a pea plant has smooth or wrinkled seeds. During meiosis, each sex cell receives only one allele governing a particular trait. Scientists refer to this as **segregation**. When a pea plant reproduces, the offspring receives one gene from the male pollen and one gene from the female part of the flower to determine its type of seeds.

Scientists now use a diagram called a **Punnett square** to predict the probability that offspring will inherit particular traits. For each allele, a capital letter is used to indicate the dominant trait, while a lower case letter is used to show the recessive trait. An organism is **homozygous** for a characteristic if both genes governing that trait are the same: for example, a pea plant may have two genes, both of which are for purple flowers. An organism is **heterozygous** for a particular characteristic if it has two different genes for that trait — one recessive and one dominant.

To the right is a Punnett square for a cross between two homozygous pea plants with purple flowers and white flowers. Notice that the offspring in this example all have the same two genes for flowers — *Pp*. This *genetic makeup* is known as the **genotype**. The *appearance* of the organism is known as the **phenotype**. In this case, a plant with the genes (*Pp*) will have purple flowers. Both (*PP*) and (*Pp*) have the same phenotype, even though they have different genotypes.

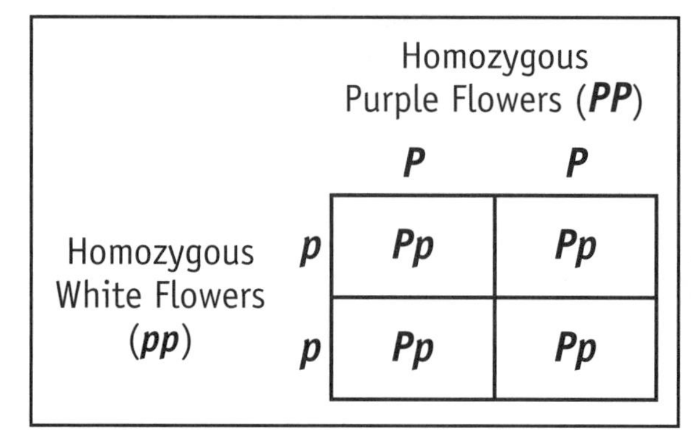

If a scientist were to self-pollinate these offspring, the results would be

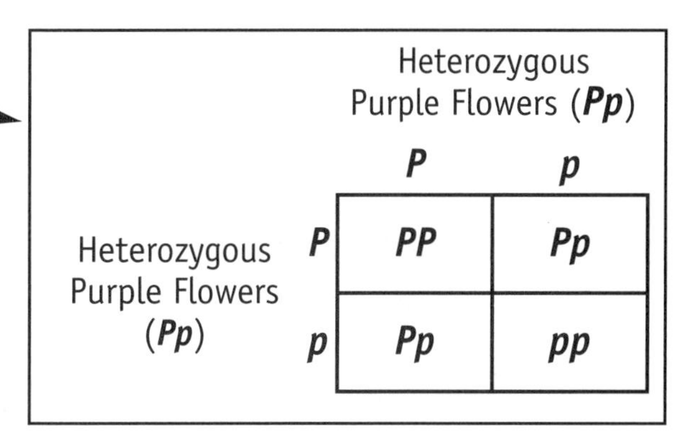

The offspring of this combination would be one half *Pp*, one quarter *PP*, and one quarter *pp*. Scientists would express the ratio of these flowers as follows: **1*PP* : 2*Pp* : 1*pp*.** In other words, for every (*PP*), there would be two (*Pp*) and one (*pp*). Because *P* is the dominant trait, three quarters of these flowers would be purple — their phenotype. The phenotype of one quarter of the flowers would be white.

In some cases, a dominant trait does not totally mask a recessive trait. Instead, both genes are potentially expressed. For example, a blue eye gene and a brown eye gene may produce green eyes. A plant with red flowers and one with white flowers may produce offspring with pink flowers. Scientists call this **incomplete dominance**.

APPLYING WHAT YOU HAVE LEARNED

✦ Construct a **Punnett square** for the offspring of two heterozygous pea plants with *wrinkled seeds* (Ww). *Smooth seeds* (w) are recessive in pea plants.

- Identify the **phenotype** and **genotype** for each potential offspring.

SEXUAL IDENTITY AND SEX-LINKED TRAITS

Have you ever wondered why some offspring are males, while others are female? One pair of chromosomes, known as the **sex chromosomes**, determine the sex of an organism.

DETERMINING SEXUAL IDENTITY

In humans, a female has two **X** chromosomes, while a male has one **X** and a shorter **Y** chromosome. During meiosis, these chromosome pairs separate. Each **gamete**, or sex cell, receives 23 chromosomes. A female egg cell always has an X chromosome. A male sperm cell, however, may have either an X chromosome or a Y chromosome. If a sperm cell with an X chromosome fertilizes an egg, the offspring is female, but if the sperm cell carries a Y chromosome, then the offspring will be male.

Notice how, in the following Punnet square, the chromosome inherited from the father (XY) determines the sex of the offspring:

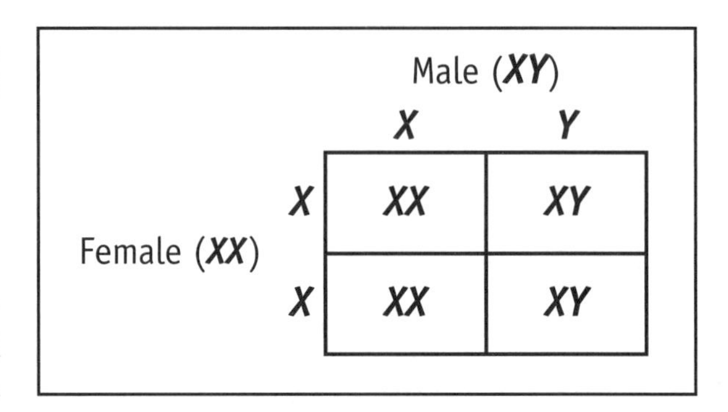

SEX-LINKED TRAITS

In addition to sexual identity, there are also **sex-linked traits**. For example, in fruit flies, red eyes are a dominant trait. The eye color gene of a fruit fly is carried on the **X** chromosome. If a female fruit fly with one red-eye gene and one white-eye gene is paired with a red-eyed male fruit fly, female offspring will inherit the dominant red-eye gene from the male and may also inherit it from the female. All of the female offspring will have red eyes. However, male offspring inherit the **Y** chromosome from the male, which has no eye color. From a heterozygous female, the male inherits either a dominant red-eye gene or a recessive white-eye gene. Half the male offspring will thus have white eyes.

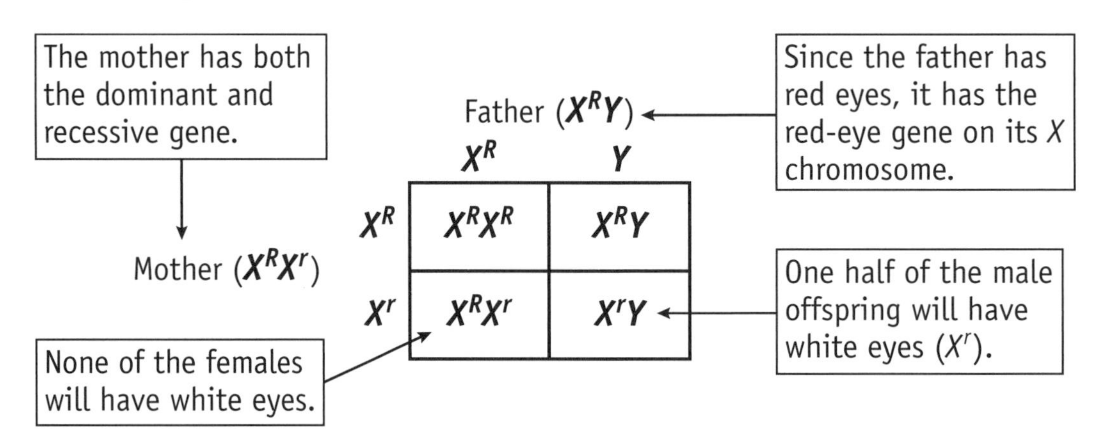

APPLYING WHAT YOU HAVE LEARNED

- Identify the phenotype for each of the four offspring in the Punnett square above.

In humans, some genetic diseases are sex-linked, such as hemophilia. Scientists sometimes trace the presence of genetic disease in a family through a diagram known as a **pedigree**. Often males are shown as squares, while females are represented as circles. A short line between a male and female indicates marriage. Lines extending from this connection indicate their children. Someone who displays the genetic disease or other sex-linked trait is filled in or colored on the chart. Someone who carries one of the disease-carrying genes but does not suffer from the disease is called a "carrier." Often, a carrier's square or circle is half filled. To answer questions on pedigree charts, first determine the genotype of the parents; then make a Punnet square to determine their offspring.Examine the following chart showing the presence of hemophilia — a sex-linked genetic disorder in which blood clotting is abnormally slow.

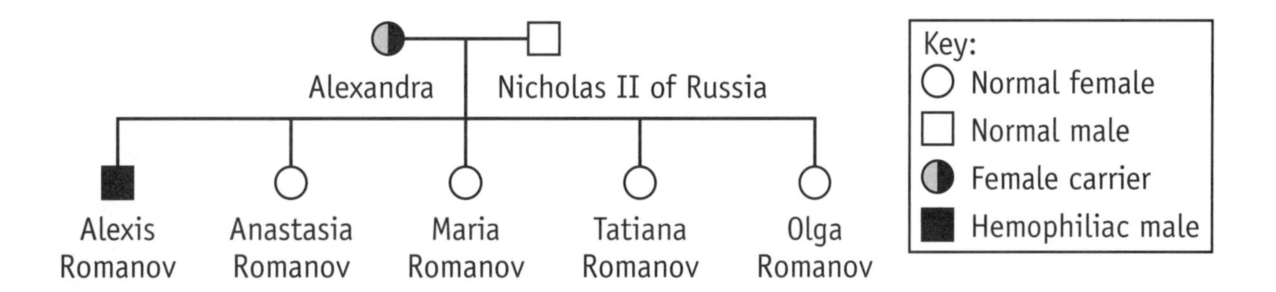

APPLYING WHAT YOU HAVE LEARNED

✦ Based on the information in this pedigree chart, if Alexandra and Nicholas had had another son, what would have been the chance that the boy would have been a hemophiliac like Alexis? Make a Punnett square to find the answer.

DNA — THE DOUBLE HELIX

The hereditary information in genes is stored in the long, twisted molecule known as **DNA**. Each of us carries billions of copies of DNA instructions in every cell of our body. Although each of our cells contains the same genes and DNA, only some of those genes are "turned on" or **expressed** in each cell. This is what allows cells to specialize their functions in multicellular organisms (*such as skin cells, nerve cells, etc.*)

THE WORK OF WATSON AND CRICK

Based on photographs of DNA crystals by Rosalind Franklin and other evidence, **Francis Crick**, a graduate student at Cambridge University, and Dr. **James Watson**, a biochemist, developed a model for the structure of DNA in 1953. They concluded that the molecule responsible for heredity, **DNA** (*deoxyribonucleic acid*), resembled a "double helix" much like a twisted ladder.

Science Photo Library

Francis Crick and James Watson (1953)

In Watson and Crick's model, the sides of the ladder are held together by steps. Each side of the DNA ladder is formed by alternating a sugar (*deoxyribose*) with a phosphate group. The "steps" of the ladder are made up of four different bases that carry the information to create an organism and to keep it running. Each base is always found in the same combination: either **Adenine** (A) with **Thymine** (T), or **Guanine** (G) with **Cytosine** (C).

The DNA of every living organism, from the simplest bacteria to human beings, is made up of these same chemical ingredients. DNA uses the combination of four bases to record its genetic information, almost like a four-letter alphabet. When the body creates a new cell, it puts together these four different "letters" of the DNA alphabet.

The structure of the DNA molecule makes it easy to **replicate** (*copy itself*). The two sides of the DNA ladder pull apart, leaving each side with one of the two base combinations. In replication, the cell "fills" in the missing parts of the DNA molecule. For example, if adenine (A) is there, it is filled in by thymine (T). At the end of the process, each half of the original DNA molecule has become a completely new DNA molecule, identical to the original DNA. The particular structure of the DNA molecule also makes it easier for other special molecules, known as **RNA**, to copy parts of the DNA chain so that they can synthesize proteins in the ribosomes.

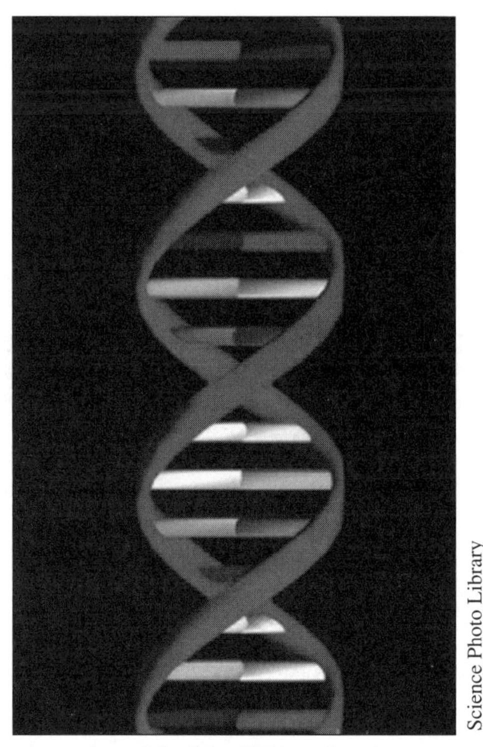

Science Photo Library

A model of the DNA molecule

MUTATION AND GENETIC ENGINEERING

Sometimes an error occurs in DNA replication. An adenine-thymine combination may take the place of a guanine-cytosine combination. Such changes in the structure of a gene are referred to as a **mutation**. Mutations may also be caused by chemicals or radiation. Some genetic mutations can cause serious disorders, like cancer. Cancer is caused by mutations that cause uncontrolled cell divisions. When mutations occur in sex cells, they are inherited by later generations. The process of mutation allows for genetic variation, a key to evolution.

Modern **biotechnology** allows scientists to examine a couple's DNA before they become parents, to prevent genetic diseases. Scientists can **splice genes** from one organism to another to repair a defective gene or reach some other result. For example, scientists have spliced the human gene for making insulin onto the DNA of bacteria to produce insulin. They can splice genes to create crops resistant to pests. The combination of DNA from two or more sources is known as **recombinant DNA**. Scientists can also use **DNA fingerprinting** to detect a person's identity, since each person has unique DNA. Scientists can even reproduce an organism from its DNA in the process of **cloning**.

In 2003, the **Human Genome Project**, an ambitious international project using DNA technology, completed mapping the 25,000 human genes — made up of 3 billion nucleotide pairs. It is hoped that the Human Genome Project will help scientists develop cures for many genetic diseases.

WHAT YOU SHOULD KNOW

★ Make sure you know that genes are packets of hereditary information made up of DNA.

★ Make sure you know that alleles are different genes that govern the same trait. In sexual organisms, an offspring gets one allele from each parent.

★ Make sure you know that dominant traits "mask" recessive traits. If one or both alleles is dominant, the **dominant trait** will appear. The **recessive trait** only appears if both alleles are recessive.

★ Make sure you know that some traits are sex-linked. They are carried by the X or Y chromosome.

CHAPTER STUDY CARDS

The History of Genetics

★ **Gregor Mendel.** Through his work with pea plants, Mendel became the first scientist to discover the laws of genetics.

★ **James Watson and Francis Crick.** They developed a model for the structure of DNA.

★ **Humane Genome Project.** A project using DNA technology that has mapped all 20,000 to 25,000 human genes.

Laws of Genetics

★ **Dominant Traits.** Mendel found that for each pair of contrasting traits, one trait was dominant; it is expressed if it is present in either gene.

★ **Recessive Traits.** Mendel found the recessive trait appeared in one of four offspring in the second generation; it is expressed only if present in both parents.

★ **Law of Independent Assortment.** Each trait is determined independently.

Gene Vocabulary

★ **Alleles.** Two genes that determine the same trait.

★ **Homozygous.** When both genes that govern a trait are the same (PP or pp).

★ **Heterozygous.** When an organism has two different genes for the same trait (Pp).

★ **Genotype.** An organism's genetic make-up for a trait.

★ **Phenotype.** The way an organism appears for a trait based on its genotype.

DNA (*Deoxyribonucleic acid*)

★ **Heredity.** DNA is the molecule responsible for heredity.

★ **Double Helix.** The shape of DNA resembles a twisted ladder, making it easier to replicate.

★ **DNA Code.** The steps of the ladder are made up of four different chemical bases. These chemical bases always appear in the same combinations:

- Adenine — Thymine
- Guanine — Cytosine

CHECKING YOUR UNDERSTANDING

1 A characteristic of pea plants is the color of the pod holding the seeds. Pea plants only have yellow pods when they possess two recessive genes (*gg*). A heterozygous genotype produces a plant with green pods. Which of the following Punnett squares shows the correct results of a cross between two heterozygous plants?

A.

Gg (top) × *Gg* (side)

	G	g
G	GG	Gg
g	Gg	gg

B.

Gg (top) × *Gg* (side)

	G	g
G	Gg	Gg
g	Gg	gg

C.

GG (top) × *Gg* (side)

	G	G
G	GG	GG
g	Gg	Gg

D.

GG (top) × *Gg* (side)

	g	g
G	Gg	Gg
g	gg	gg

This question tests your understanding of genetics. You must interpret a Punnett square and know the meaning of "heterozygous." An organism is heterozygous for a particular characteristic if it has two different genes for that trait — one recessive and one dominant. Which answer correctly shows heterozygous parents and their offspring?

Now try answering some additional questions on your own dealing with topics covered in this chapter — heredity, genetics, and DNA.

2 Which statement describes the work of Gregor Mendel?

- ♦ Examine the Question
- ♦ Recall What You Know
- ♦ Apply What You Know

A. He developed basic principles of heredity without having knowledge of chromosomes.

B. He explained the principle of dominance on the basis of the gene-chromosome theory.

C. He developed the microscope for the study of genes in pea plants.

D. He used his knowledge of gene mutations to explain the appearance of new traits in organisms.

LS: C 10–22

3 **In canaries, the gene for singing (*S*) is dominant over the gene for non-singing (*s*). When heterozygous singing canaries (Ss) are mated with non-singing canaries, what percentage of their offspring is likely to possess the singing trait ?**

A. none
B. 25%
C. 50%
D. 100%

LS: C 10–8

4 **In humans, the gene for polydactyly (*extra fingers or toes*) is dominant over the gene for the normal number of digits. If parents who are both homozygous dominant for polydactyly (PP) have four children, how many of the children would probably have extra fingers or toes?**

A. 0
B. 2
C. 3
D. 4

LS: C 10–8

5 **The instructions for the various traits of an organism are found coded in the arrangement of**

A. glucose units in its carbohydrate molecules
B. the bases in DNA molecules in the nucleus
C. molecules in its cell membrane
D. energy-rich bonds in its ATP molecules

- ♦ Examine the Question
- ♦ Recall What You Know
- ♦ Apply What You Know

LS: C 10–8

Use the information in the chart below to answer the following two questions.

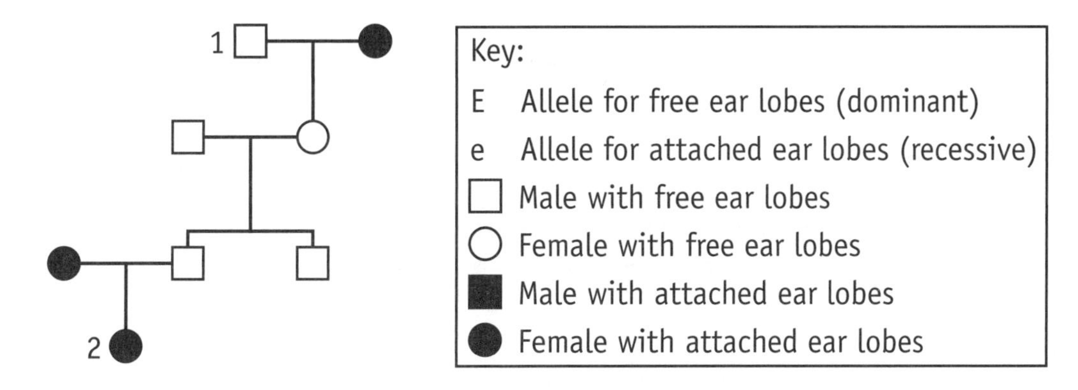

6 **The chart shows a history of ear lobe shapes. The genotype of individual 1 could be**

A. EE only
B. Ee only
C. ee only
D. EE or Ee

LS: C 10–8

7 **According to the chart, the genotype of individual 2 could be**

A. EE only
B. Ee only
C. ee only
D. EE or Ee

LS: C 10–8

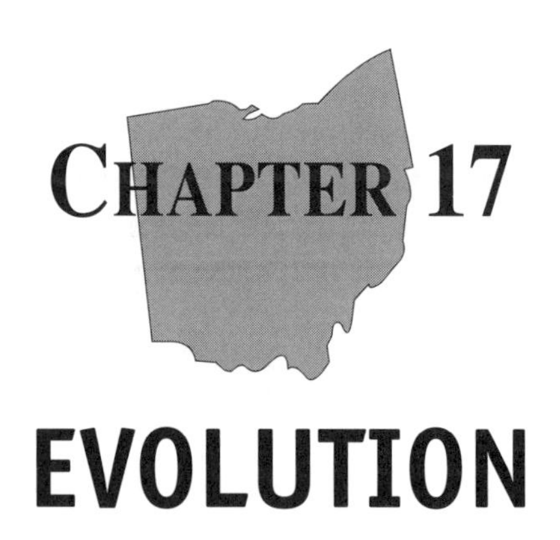

CHAPTER 17

EVOLUTION

In this chapter, you will learn about **biological evolution** — the process through which successive generations of living things gradually change and adapt to their environment.

MAJOR IDEAS

A. Biological evolution refers to gradual changes in a species over successive generations, based on changes in gene frequency. Through evolution, a species is able to adapt to changes in its environment.

B. Charles Darwin explained biological evolution through natural selection. Random inherited differences give some members of a species an advantage in surviving and reproducing. The proportion of individuals in a species with advantageous characteristics will gradually increase. Several different types of evidence support Darwin's theory. These include links between fossils and living organisms, the existence of vestigial structures, and similarities in related organisms' DNA and embryonic development.

C. Evolutionary processes have led to a great diversity of forms of life on Earth. Scientists use different systems of biological classification to understand this diversity and how different life forms are related.

THE BIRTH OF EVOLUTIONARY THEORY

Several developments led to the theory of evolution in the mid-nineteenth century. First, scientists were already familiar with **breeding** — the practice of farmers in selectively using certain plants and animals to produce offspring. Through breeding, farmers had developed improved strains of wheat and meatier cattle. Secondly, nineteenth-century geologists discovered that Earth was much older than previously thought. This gave plants and animals the time they needed to evolve. Geologists also became better at identifying fossils in the sedimentary rock of Earth's crust. Finally, exploration of the world's different regions had led naturalists to record differences and similarities between various species of plants and animals.

JEAN BAPTISTE DE LAMARCK

Using the fossil record, **Jean Baptiste de Lamarck** (1744–1829), a French scientist, proposed that different species with similar characteristics descended from a common ancestor. Underlying his theory was the belief that organisms are in a constant state of advancement. Each successive generation constantly "improves" on its predecessors. This improvement occurs too slowly to be seen, but was detectable by examining fossil records. Lamarck saw humans at the top of this pyramid of progression. Based on his study of related birds, Lamarck concluded that organisms made certain structural changes to adapt to their environment, and then passed these changes on to their offspring. His idea that an organism could pass "acquired traits" on to children was rejected by other scientists.

Jean Baptiste de Lamarck

CHARLES DARWIN

A new theory of evolution was proposed by **Charles Darwin**. In 1859, he published *The Origin of Species*. He based his work on his prior travels as the naturalist on a British ship. During his travels to the Galapagos Islands, Darwin was struck by the wide variety of finches with different beaks. Darwin found that each type of beak was suited to using a different food source. Like Lamarck, he reasoned these birds were descended from a common ancestor, which had flown to the islands from South America. But instead of arguing, as Lamarck had done, that the birds passed on acquired traits to their offspring, Darwin came up with the **theory of natural selection**.

Charles Darwin

According to Darwin, there were always random variations within a species. Each individual organism always inherited slightly different characteristics than the other members of its species. Although these differences are random, they give certain members of the species an advantage over other members in their ability to survive and reproduce. Because the natural environment may not have enough food, or other resources — and because some individuals will be killed by other organisms — only a proportion of each species will live to adulthood and reproduce. Favorable characteristics are those that help an organism to cope with its environment. More individuals with favorable characteristics would be able to survive and reproduce, so the proportion of individuals within the species with these hereditary characteristics would gradually increase. Darwin called this "the survival of the fittest" through natural selection.

Darwin was unable to explain the reason for the random variations among the members of a species, but scientists now believe the cause of these differences is **genetic mutation**. Some of these mutations give rise to inherited differences in traits.

APPLYING WHAT YOU HAVE LEARNED

- Compare Lamarck and Darwin's theories. (1) How are they similar? (2) How are they different? (3) How does this comparison demonstrate that different scientists may contribute to the growth of a scientific idea?

THE EVIDENCE FOR EVOLUTION

Scientists today believe there are several sources of evidence supporting evolutionary theory.

THE FOSSIL RECORD

Most fossils are found in sedimentary rock, which is formed when layers of sand, dust, soil, and mud are deposited on Earth's surface and compacted. **Fossils** are impressions left by life forms in sedimentary rock. When a plant or animal dies, it decays and is buried under layers of mud or sand that become sedimentary rock. The organism may leave an imprint or its hard parts may be replaced by minerals which form the fossil. The fossils of more recent organisms are closer to the surface; older fossils are buried in deeper layers. Geologists now use radiocarbon dating to tell the approximate age of a fossil. The fossil record suggests that current life forms are related to older life forms, and that organisms have evolved slowly over time.

VESTIGIAL STRUCTURES

Some organisms contain organs that serve no useful function, like the human appendix or the hind leg bones of a whale. Such vestigial structures suggest that these organisms evolved from others that once used these structures. Vestigial structures provide a strong argument for the idea of a common ancestry among members of any group that share those structures.

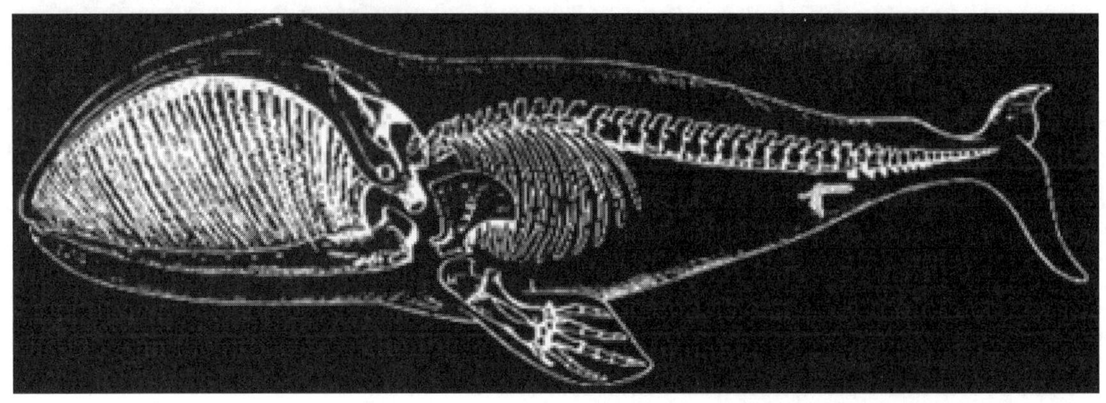

Skeleton of a whale, showing vestigial pelvic bones

SIMILARITIES IN EMBRYONIC DEVELOPMENT

Most **vertebrates** (*animals with backbones*) have a similar appearance as early **embryos** (*organisms before birth*). Such similarities once again suggest common origins.

DNA EVIDENCE

Scientists have found that different species of organisms that appear to be closely related in fact have similarities in their DNA. This powerful evidence again suggests that these organisms have evolved from a common ancestor.

OBSERVATION OF ADAPTATIONS

Scientists have observed how selective breeding by farmers and other breeders have led to changes within a species such as dogs. They also see how a population of insects or microorganisms is able to evolve so that they can resist an insecticide or antibiotic. For example, most of an insect population may be killed by an insecticide, but the survivors are able to pass on genes to their offspring enabling them to resist the insecticide. Such evidence further supports the theory of evolution through natural selection.

APPLYING WHAT YOU HAVE LEARNED

- ✦ Explain two types of evidence that help to support evolutionary theory.

THE EVOLUTION OF LIFE ON EARTH

Scientists now believe that Earth is about 4.6 billion years old. When Earth first formed, its atmosphere was different from today. Scientists differ on its exact composition, but laboratory experiments suggest that simple organic compounds may have formed from Earth's primitive atmosphere, and that some of these compounds were able to form cell-like structures with protein membranes known as **microspheres** and **coacervates**. Unlike cells, however, these microsphere structures lacked DNA. Many scientists also believe that self-replicating RNA may have developed independently and eventually joined with the cell-like microspheres.

ARCHAEBACTERIA

Bacteria are the oldest known forms of life and have populated Earth for most of the time our planet has existed. The first primitive cells probably emerged about 4 billion years ago as unicelled prokaryotes similar to modern **archaebacteria**. These organisms obtained energy from chemical processes (*chemosynthesis*) rather than photosynthesis.

The oldest known fossils, about 3.5 billion years old, were left by prokaryotic cells capable of photosynthesis. These photosynthetic cells gradually added oxygen to the Earth's atmosphere. Between one and two billion years ago, the first eukaryotic cells seem to have emerged.

Scientists now hypothesize that small aerobic prokaryotes may have entered larger prokaryotic cells. These smaller prokaryotes now make up the mitochondria and chloroplasts of modern cells. This may be why mitochondria and chloroplasts have double membranes. With the development of eukaryotic cells, complex multicellular organisms eventually evolved.

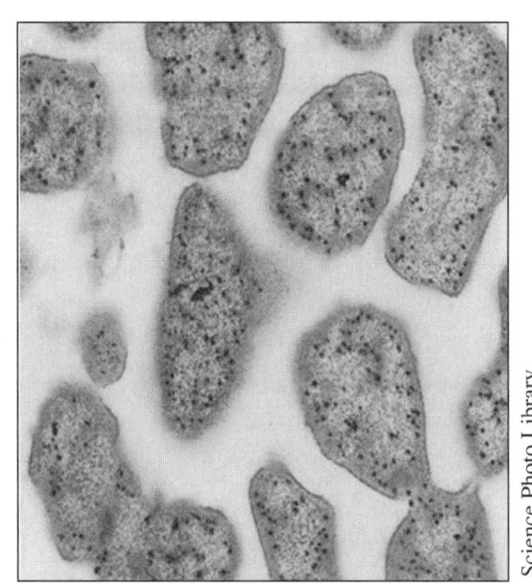

Science Photo Library

Archaebacteria

THE DIVERSITY OF LIFE

The processes of natural selection, described earlier, have contributed to the development and diversification of more complex life forms. For example, when an environment changes, the survival value of the inherited characteristics of an organism may change. Thus, rabbits with white fur in winter may survive better in colder climates than other rabbits. If the climate gets colder, rabbits with this characteristic will be more likely to survive and produce more offspring.

Gradually, the genetic make-up of organisms changes and diversifies. Evolution can therefore be viewed as being based on changes in **gene frequency** — the percentage of particular genes within a population. Although evolutionary mechanisms have given rise to a vast variety of organisms on Earth, they also indicate that many or all of these forms of life share common ancestors — highlighting the unity as well as the diversity of life.

Ducks and flamingos are quite different, yet they probably share a common ancestor.

OTHER FACTORS IN EVOLUTION

In addition to natural selection, several other evolutionary mechanisms have contributed to the wide diversity found among the forms of life on Earth:

Genetic drift. This refers to random events in small populations, like the failure of some organisms to reproduce, which can act with natural selection to quickly change the characteristics of that group of organisms over time.

Migration. A population may also change because of emigration or immigration. **Emigration** refers to the movement of organisms out of a group, while **immigration** is the movement of organisms into the group.

Mating. Mating practices in a group can also affect its development and evolution. In many groups, mating is not random. Individuals with certain characteristics are more often selected as mates. This affects which genes from the "gene pool" are passed on to the population's offspring.

Speciation. A **species** is a group of organisms with a similar structure and appearance who can mate and produce fertile offspring. **Speciation** refers to the creation of a new species. Evolutionary developments lead to speciation. If, for example, a species is geographically divided in two by the creation of a new river or mountain range, the two groups may not have any opportunity to breed together. Each group then continues its own evolutionary development independently. Eventually, the two groups become so different they can no longer interbreed, and each group becomes a new species. Even when a species is not divided geographically, members with particular traits may restrict their mating to others with similar traits. Eventually, two or more different groups develop and become separate species.

Adaptive Radiation. In the process known as **adaptive radiation**, individual organisms adapt to the environment in different ways. This leads to several subgroups, similar to the different types of finches Darwin saw on the Galapagos Islands. He noticed that each type of finch differed in beak shape and size, depending upon the food they ate. Darwin believed that these different types of finches, coming from a common ancestor, had successfully adapted to their environment in varied ways by natural selection.

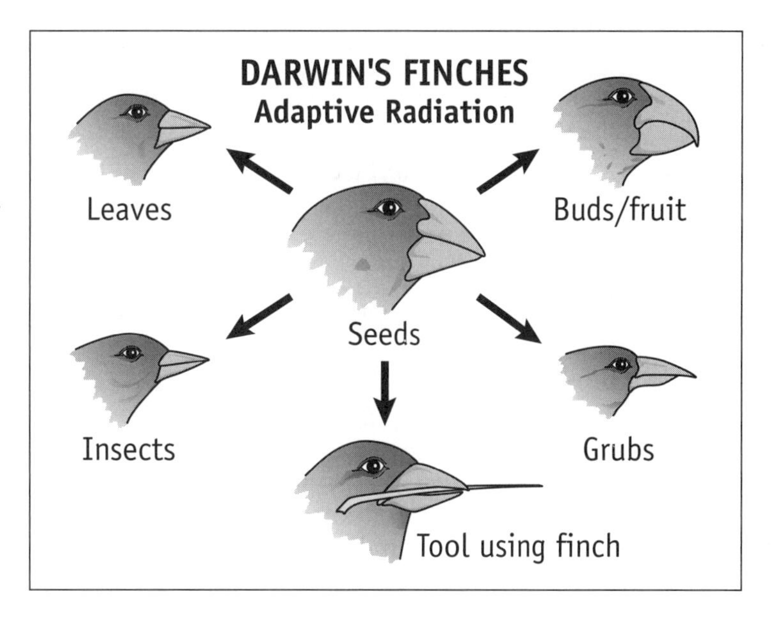

None of the mechanisms leading to evolutionary change occur overnight. Instead, these processes may take millions of years. Species also sometimes die out. When the last member of a species dies, the species becomes **extinct**. Often, a species becomes extinct because of changes in the environment. The members of the species have difficulties adapting to the challenges of environmental change. In the history of Earth, there are several periods when large numbers of species became extinct at the same time. One of these periods occurred about 350 million years ago, when large numbers of amphibians became extinct. A second period occurred about 65 million years ago, when dinosaurs and many other species became extinct. Scientists believe these mass extinctions may have occurred because of changes in the atmosphere caused by dust from a large meteor or asteroid that crashed into Earth.

Science Photo Library

BIOLOGICAL CLASSIFICATION

To understand the diversity of life, scientists use a system of **biological classification**. To **classify** is to place things into similar groups. Although organisms can be arranged in a variety of ways, most biologists arrange them into a hierarchy of groups and subgroups based on similarities and differences that reflect their evolutionary relationships. Under the most common system of classification, all organisms are divided into six "kingdoms":

Kingdom	Description
Archaebacteria	Prokaryotic, unicellular organisms that are anaerobic — live without the use of oxygen.
Eubacteria	Prokaryotic, unicellular organisms that generally use oxygen.
Protista	Unicellular and simple multicellular eukaryotic organisms, like amoeba, paramecia, and algae.
Fungi	Unicellular and multicellular organisms similar to plants but without photosynthesis, such as mushrooms.
Plants	Multicellular organisms that use photosynthesis and have cell walls.
Animals	Multicellular organisms that can move around, have nervous systems, and lack cell walls or photosynthesis.

The highest category in most classification systems is the **kingdom**. Each kingdom is further divided into groups. Immediately below the kingdom is the **phylum**. Organisms in each kingdom are divided into **phyla** based on structural similarities. For example, crabs, lobsters, insects, and spiders are all **anthropods**, since they all have external skeletons and jointed bodies and limbs. Each phylum is then divided into *classes*, *orders*, *families*, *genera*, and *species*. The **species** is the fundamental unit in the classification of organisms.

Science Photo Library

Anthropoda

Genus and Species. The scientific name for an organism consists of its genus and its species. When scientists name an organism, the first word is the genus name, and the second word denotes the species. The genus is always capitalized.

Common Name	Genus	Species
Human being	*Homo*	*sapiens*
House cat	*Felis*	*domesticus*

WHAT YOU SHOULD KNOW

★ You should know that the theory of evolution was developed in the nineteenth century. Lamarck believed evolution occurred through "acquired traits." Darwin introduced the concept of "natural selection" and "survival of the fittest."

★ Make sure you know that scientists now explain biological evolution through natural selection and genetic mutation.

★ Evolution is based on changes in the gene frequency of a species over time. Migration, genetic drift, environmental change, and adaptive radiation also play a role in evolution.

★ Life on Earth began with unicellular microorganisms about 4 billions years ago, and later evolved into increasingly complex multicellular organisms.

★ Scientists use different systems of biological classification to show how various forms of life are related. The *species* is the fundamental form of classification.

CHAPTER STUDY CARDS

Leading Theorists in Evolution

★ **Jean Baptiste Lamarck.** Believed organisms made changes to adapt to their environment, and passed these "acquired traits' on to their offspring.

★ **Charles Darwin.** His **theory of natural selection** argued that individuals in a species inherit slightly different traits, giving certain members an advantage over other members of the species to survive and reproduce. Those with traits best fitted to their environment produce more offspring, until all of the species have those traits.

The Evidence for Evolution

★ **Fossil Record.** Current life forms are related to older ones, evolving over time.

★ **Vestigial Structures.** Supports the view that a common ancestry exists among members with similar structures.

★ **Similarities in Embryo Development.** Most vertebrates have a similar appearance, suggesting they had a common origin.

★ **DNA Evidence.** Organisms that have similar DNA had common ancestors.

★ **Observation of Adaptations.** Indicates that organisms learn to adapt to changes.

Mechanisms That Impact the Diversity of Life

★ **Genetic Drift.** Random events affect the gene pool of small populations.

★ **Emigration/Immigration.** Movement of organisms out of or into a group.

★ **Mating.** Individuals with certain traits are picked more often as mates, affecting which genes are passed on.

★ **Speciation.** Process that leads to the creation of a new species.

★ **Adaptive Radiation.** Species diversify as they adapt in different ways.

Biological Classification

★ **Kingdom.** This is the highest category in the most common classification system.

★ **Phylum.** Organisms are grouped based on similarities in body plan and organization.

★ **Classes, Orders, Families, Genera.** The next levels of classification.

★ **Species.** The fundamental unit in the classification of organisms.

Genus and Species. The way scientists name an organism. The genus name and then the species: e.g. *Homo sapiens.*

CHECKING YOUR UNDERSTANDING

Use the information in the chart on the next page to answer the following question.

1 Organism X appeared on Earth much earlier than organism Y. Many scientists believe organism X appeared between 3 and 4 billion years ago, while organism Y first appeared approximately 1 billion years ago. Based on the chart, which most likely describes the sequence of life on earth?

A. (A)

B. (B)

C. (C)

D. (D)

◆ Examine the Question
◆ Recall What You Know
◆ Apply What You Know

LS: E 10–25

	Organism X	Organism Y
(A)	simple multicellular	unicellular
(B)	complex multicellular	simple multicellular
(C)	unicellular	simple multicellular
(D)	complex multicellular	unicellular

This question examines your understanding of the evolution of life on Earth. Recall that the first life to appear on Earth consisted of simple unicelled organisms (*archaebacteria*). From these earliest life forms, more complex organisms later evolved. Choice C best reflects this information.

Now try answering some questions on your own dealing with the topics covered in this chapter — genetic mutation and variation, natural selection; and the evolution and classification of life forms.

2 A man lifts weights and develops large arm muscles. His son has larger muscles than his father had at the same age. According to Lamarck's theory, this situation would be due to

LS: J 10–22

A. competition between father and son
B. survival of the fittest
C. inheritance of acquired characteristics
D. mutation of genes

3 The process of biological "evolution" is best described as

LS: H 10–20

A. a process of change in the frequency of genes in a population over time
B. a process by which organisms become extinct over time
C. the reproductive isolation of members of certain species
D. the replacement of one community by another

4 A large population of houseflies was sprayed with a newly developed, fast-acting insecticide. The appearance of increasing numbers of houseflies that are resistant to this insecticide supports the view that

LS: I 10–23

A. a species' traits tend to remain constant
B. the environment does not change
C. genetic variation permits species adaptation
D. house flies are a form of anaerobic archaebacteria

Use the information in the diagram to answer the following question.

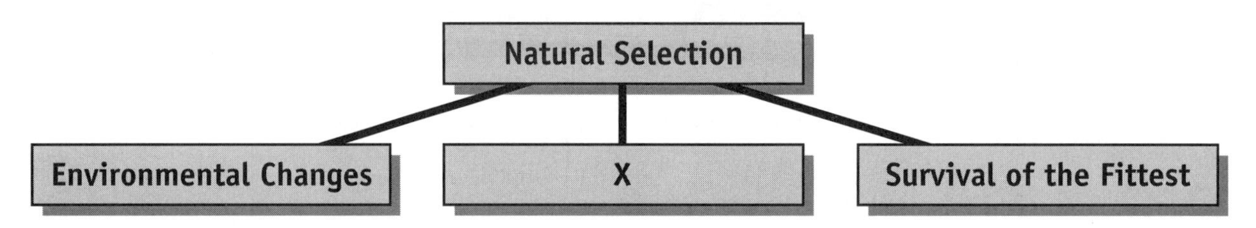

5 Some of the concepts included in Charles Darwin's theory of natural selection are represented in the diagram above. Which of the following concepts would be correctly placed in box X?

A. random inherited differences
B. vestigial structures
C. incomplete dominance
D. transmission of acquired traits

LS: I 10–21

Use the information below to answer questions. 6–8.

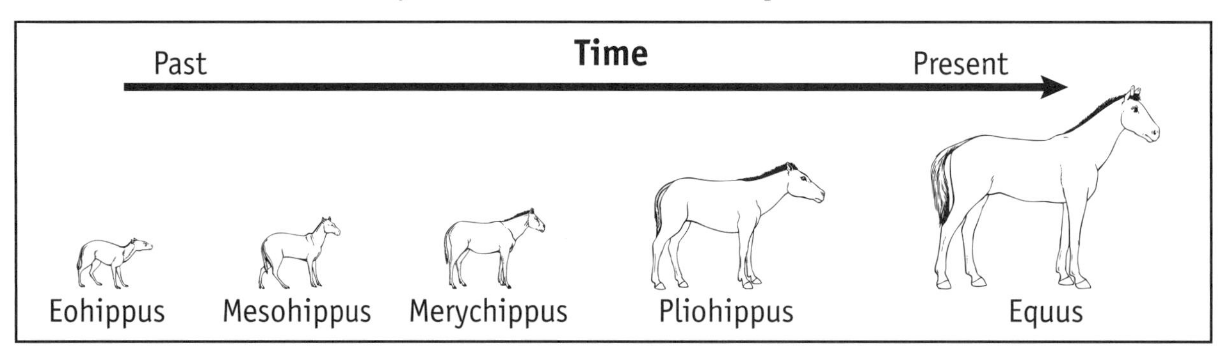

6 The diagram above shows the gradual changes taking place over time in the structural anatomy of the present-day horse (Equus). Which concept is illustrated by the physical variations in the horse as its body size and structure changed over time?

A. law of independent assortment
B. incomplete dominance
C. binary fission
D. theory of natural selection

LS: I 10–21

7 How did scientists obtain the information needed to construct this diagram?

A. They examined horses in different biomes.
B. They analyzed various fossils.
C. They made projections from DNA evidence.
D. They investigated changes in Earth's climate.

LS: I 10–23

8 What conclusion can be drawn from the diagram?

A. Horses became larger to escape predators.
B. Horses became larger to eat leaves from shrubs and bushes.
C. Horses became smaller to make it harder for predators to see them.
D. Horses became smaller to make it easier for them to eat insects.

LS: I 10–21

CHAPTER 18

ECOLOGY

All organisms depend on both the environment and other organisms to survive. **Ecology** is the study of relationships between living organisms and their environment. Ecologists often study how different organisms live and interact in a specific area. All of the living organisms in an area, together with their nonliving environment, make up what ecologists call an **ecosystem** (*ecological system*).

Ohio's Mentor Marsh Nature Preserve forms an ecosystem.

The size of an ecosystem can vary greatly from a small pond to a vast forest. Different ecosystems are often separated by geographical barriers, such as deserts, mountains or oceans. However, the borders of ecosystems are not rigid, and one ecosystem often blends into another ecosystem. An ecosystem, such as a tropical rainforest, may also have several smaller ecosystems within it, such as a forest canopy ecosystem and a forest floor ecosystem.

MAJOR IDEAS

A. The organisms and environment of an ecosystem are **interdependent**.

B. An ecosystem always includes both **biotic components** (*living things*) and **abiotic components** (*nonliving things or conditions, such as temperature, water, sunlight, and minerals*).

C. Ecosystems recycle both matter and energy.

D. Ecosystems fluctuate around a state of equilibrium or balance. Drastic events, such as a fire or climatic change, may trigger changes to an ecosystem.

E. Human activities now threaten many of the world's ecosystems.

ECOSYSTEMS

There are many types of ecosystems. Ecosystems on land are known as **terrestrial ecosystems**. These include forest, grassland, desert, taiga and tundra ecosystems. Ecosystems in water are known as **aquatic ecosystems**. These include freshwater ecosystems like the Great Lakes, and ocean zone ecosystems like the Continental Shelf, Great Barrier Reef, and the mid-ocean floor.

Community and Population in an Ecosystem. All the living organisms of different species found in a single ecosystem are referred to as a community. The **community** is made up of all the bacteria, fungi, plants and animals of the ecosystem. For example, trees, squirrels, foxes, and deer make up the community of a temperate forest ecosystem. All of the organisms of the same species in a particular ecosystem are known as the **population** of that species. For example, all of the small mouth bass found in the Mississippi River Basin make up that ecosystem's *small mouth bass population.*

INTERACTION WITH ABIOTIC COMPONENTS

Many **abiotic** (*nonliving*) factors affect the living organisms in either an aquatic or terrestrial ecosystem. These factors include temperature, the availability of water, humidity, the soil, minerals, the amount of sunlight, and the concentration of oxygen and other gases. Since temperatures and other abiotic factors often vary, organisms usually have a range of conditions they can *tolerate*. Many organisms, like mammals, use their own energy to maintain similar internal conditions despite changes in their physical environment. Other organisms, like reptiles, change some of their internal conditions (*such as body temperature*) in response to changes in the external environment.

The iguana is sensitive to its environment.

Species sometimes use specific ways to avoid or reduce changes in their physical environment. For example, many desert species sleep during the day and come out in cooler night temperatures. Other animals hide underground in cold weather or migrate southwards each winter to places that are warmer.

Drastic changes in the environment often lead to a **succession** of changes in an ecosystem. For example, a fire may destroy ancient trees in a temperate forest ecosystem. Weeds and grasses will quickly spring up in the ashes. Later, shrubs and taller plants will grow, blocking the light for the grasses. Pine trees and finally new deciduous trees (*trees that shed their leaves in the fall*) will take root, crowding out the shrubs. A severe change in climate can similarly alter the balance of an ecosystem and trigger a sucession of changes.

APPLYING WHAT YOU HAVE LEARNED

- ✦ Identify two types of ecosystems. What is the difference?
- ✦ Identify four abiotic components in a pond ecosystem.

INTERACTION BIOTIC COMPONENTS

Biotic (*living*) factors also affect ecosystems. All animals and plants require resources — nutrients, energy, and space — to survive. Many of these resources are provided by other living things. Every ecosystem therefore contains different populations of species that interact together. There are five ways that different species in an ecosystem generally interact:

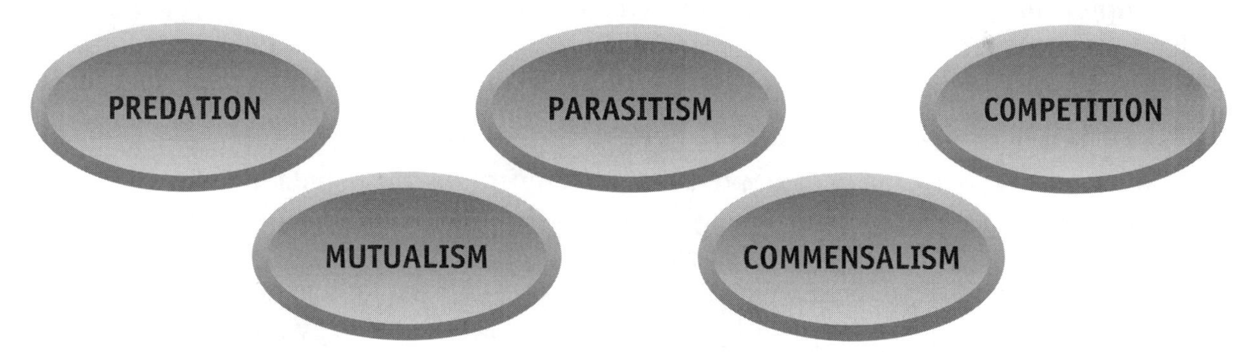

Predation. In predation, one organism — the **predator** — captures and kills another organism, known as the **prey**, for food. Through natural selection, predators like wolves and tigers developed special characteristics like sharp teeth, eyes in the front of the head, and the ability to run fast in order to hunt their prey. Through natural selection, prey also developed special physical characteristics like camouflage or eyes on the sides of the head to avoid being caught. Predators sometimes provide a useful check on their prey, which might otherwise crowd out other species through competition.

Parasitism. A **parasite** feeds on another organism — the **host**. Unlike predators, the parasite does not kill the host, at least not immediately. Some parasites, such as ticks, attach themselves on the surface of the host. Others, known as **endoparasites**, live inside the host. Endoparasites include some types of bacteria, protozoa, and tapeworms.

Competition. When two or more species in an ecosystem share similar characteristics and require the same resources, they compete. Because there is not enough of those resources for both species to multiply unchecked, either one of these species will eliminate the other, or natural selection will lead to the development of differences among the competitors. This is precisely what Darwin observed among the finches of the Galapagos Islands. On some islands, different groups of finches had developed unique features so that they would not compete for the same food. Some species of finches had large beaks to crack and eat seeds, while others had narrow beaks for searching out insects between rocks. These differences reduced **competition** among the birds.

Sometimes when a new species is introduced to an ecosystem from somewhere else, this has drastic effects. Because it is **alien**, the new species may not face predators or parasites. Its population is able to grow unchecked, while it competes for resources previously used by native species. For example, ships brought zebra mussels from Europe to the Great Lakes in the 1980s. They spread rapidly, and now compete with native species for scarce resources.

Mutualism. Mutualism refers to a cooperative relationship in which two or more species mutually benefit. For example, flowering plants provide nectar to bees and other insects. The insects then carry pollen from one plant to the flowers of other plants, leading to cross-fertilization. Both the plants and the insects *mutually* benefit from this relationship.

Commensalism. This refers to a one-sided relationship in which one species benefits without harming the other species.

EQUILIBRIUM WITHIN AN ECOSYSTEM

With unlimited resources and without predation or competition, the population of each species in an ecosystem would experience rapid growth. However, the existence of limited resources (*water, organic molecules, oxygen, carbon dioxide, space*), competition with other species and the loss of individuals to predators poses a severe check on the growth of each species. The population of a species that an area will generally support is referred to by ecologists as its **carrying capacity**. There may be cyclical fluctuations within the community of an ecosystem, but a delicate **equilibrium** (*balance*) is usually reached among its various species. This equilibrium can be upset by drastic natural disasters, climate changes, the introduction of new species, or the disappearance of an existing species. Such changes can have effects that ripple through the ecosystem. For example, predators like wolves may keep the number of deer in check. Human efforts to reduce or eliminate the wolves could have unforeseen effects. Without the wolves, the deer population might expand too rapidly, threatening plant life and competing plant-eating animals in the same ecosystem.

APPLYING WHAT YOU HAVE LEARNED

✦ Provide **one** example of your own for each of the following types of relationships: (1) predation; (2) parasitism; (3) competition; (4) mutualism; and (5) commensalism.

THE FLOW OF ENERGY AND NUTRIENTS THROUGH AN ECOSYSTEM

Energy and matter — in the form of water, nutrients, and waste — are continually recycled within an ecosystem. Ecologists can trace this flow of energy and nutrients through an ecosystem.

AUTOTROPHS (PRODUCERS)

The basic source of energy for most ecosystems is sunlight. **Producers** capture this energy and turn it into carbohydrates through photosynthesis. In terrestrial ecosystems, the main producers are plants. Plants use sunlight in photosynthesis, but they also require carbon dioxide, water, and nutrients from the soil. In aquatic ecosystems, microscopic phytoplankton use photosynthesis. Because producers make their own food, they are also known as **autotrophs**.

HETEROTROPHS (CONSUMERS)

Heterotrophs cannot make their own food. They are **consumers** that eat other organisms or their wastes to obtain organic molecules containing energy.

- ★ **Herbivores** eat plants or phytoplankton.
- ★ **Carnivores**, such as tigers, eat other animals.
- ★ **Omnivores**, such as dogs or humans, eat both plants and animals.

DECOMPOSERS

Some organisms, like vultures, ants, and bacteria, consume dead organisms or animal wastes. Decomposers, such as worms, bacteria and fungi, break down dead organisms and wastes into organic molecules. These organic molecules, especially nitrates, then become part of the soil where they are used by plants.

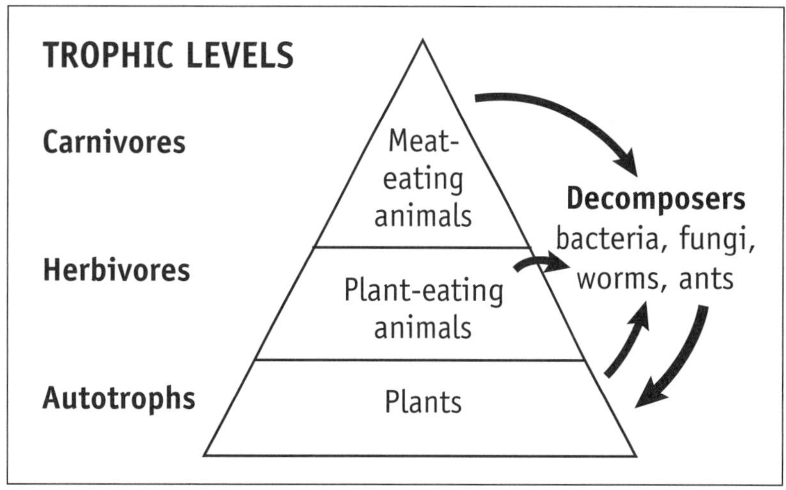

*Each level of organism eating the same kind of food in an ecosystem is referred to as a **trophic level**.*

A **food chain** or **food web** is a diagram tracing the flow of energy and nutrients through a single ecosystem. It shows the specific links between organisms in the ecosystem. The arrows indicate the direction in which energy and nutrients flow. For example, the grasses (*producers*), rabbits (*herbivores*) and coyotes (*carnivores*) on a prairie form a single ecosystem. Energy flows from the sun to the grasses. Rabbits feed on the grasses and absorb their energy and nutrients. Coyotes eat some of the rabbits and absorb their energy and nutrients. When the rabbits and coyotes die, worms, bacteria, and fungi decompose their bodies and enrich the soil for the grasses.

THE FOOD CHAIN IN A PRAIRIE ECOSYSTEM

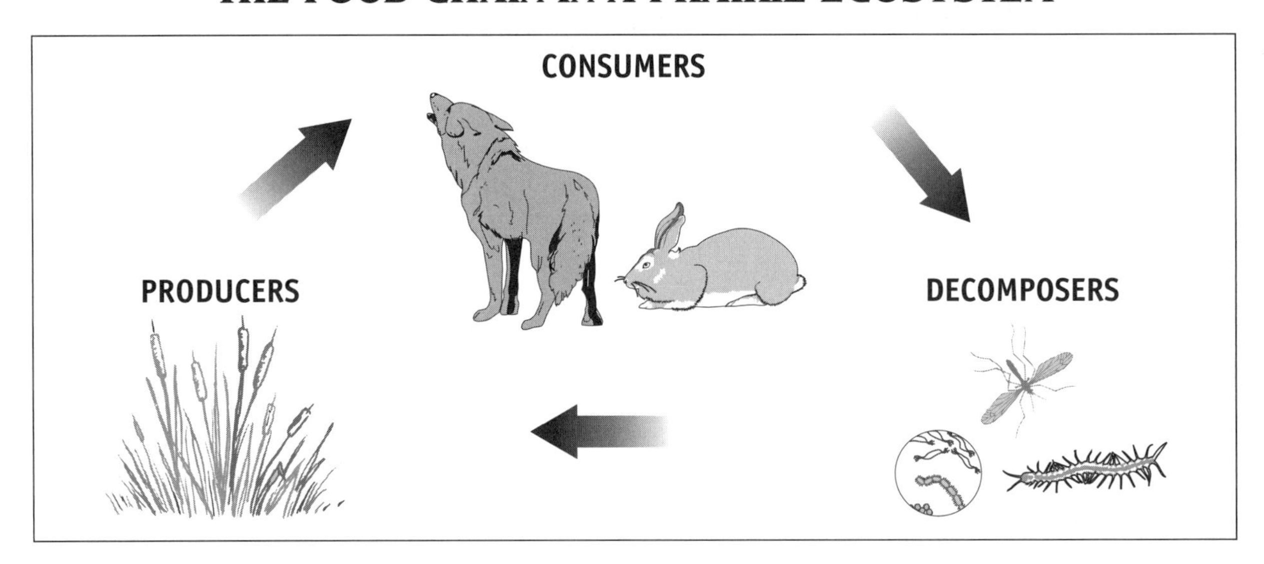

Water, carbon dioxide, and nitrates are similarly recycled through an ecosystem. For example, water is stored in the ocean, freshwater sources, or underground. It evaporates into the atmosphere, where it becomes water vapor. When the atmosphere is saturated, the water comes back to Earth's surface as rain, snow, sleet or fog. Plants take this water through their roots, and animals drink water. All organisms release water back into the environment.

Only a limited amount of energy can be transferred from one **trophic level** to the next higher level. Ecologists estimate that on average about one-tenth of the energy in a trophic level is used by the next level. The herbivores in a prairie ecosystem may consume as much as one-fifth of the energy stored by the ecosystem's grasses and other plants. Carnivores like coyotes then consume less than one-tenth of the energy found in the herbivores. Energy that is not so transferred is used by organisms at their own trophic level. Much of this energy is used to maintain the body temperature or other metabolic functions of each organism, and then escapes into the atmosphere as thermal energy (*heat*).

APPLYING WHAT YOU HAVE LEARNED

✦ Draw a food chain or web showing yourself as one of the consumers. Include in your food chain what you ate for dinner last night, and consider where the energy in those ingredients came from.

CASE STUDY: THE GREAT LAKES ECOSYSTEM

The **Great Lakes ecosystem** includes eight states in the United States and two provinces in Canada. About 33 million people live in this ecosystem, covering the five Great Lakes and its adjoining areas. It contains many smaller ecosystems.

Abiotic Components. The Great Lakes ecosystem is located where the forests of the eastern United States, the pine forests of Canada and the grassy prairies of the Midwest meet. In the southwest of the region are the flat lands of the prairies. Because the prairies are drier and flatter, wind and occasional fires keep forests from invading this area. The Great Lakes dominate this ecosystem and influence its climate. One-fifth of all the world's fresh surface water is found here. Rainfall passes into streams and rivers which drain into the lakes. Much of this water then evaporates from the Great Lakes back into the air.

Biotic Components. The Great Lakes ecosystem contains a wide variety of plant and animal life. The northern areas of the region are home to large pine forests. Many of the species of fish in the lakes are unique. Along the shores of the lakes are some the world's largest freshwater sand dunes. Plants and animals have adapted to this unique environment by developing new kinds of species. For example, the Lake Huron locust is found only in this area.

Human-Environmental Interaction. Humans have had a tremendous impact on the Great Lakes ecosystem. Some of their activities have threatened the ecosystem. For example, in the 1800s, forests were cut down to make lumber and furniture. Eventually, logging companies planted new trees. People also once threatened the fish of the Great Lakes by over-fishing and introducing alien fish that competed with native species. In more recent times, the main threat to the Great Lakes ecosystem has come from pollution. Companies and cities dumped chemicals into the atmosphere and lake waters. Many dumpsites in inland areas contain dangerous chemicals. Efforts to curb pollution may save this ecosystem.

Library of Congress

Factories sometimes pour smoke, fumes and chemicals into the air, polluting the environment.

THE IMPACT OF HUMAN ACTIVITY ON THE WORLD'S ECOSYSTEMS

Because each of the world's ecosystems represents a delicate balance, many ecosystems are threatened today by the rising human population and changes in technology.

ENVIRONMENTAL PROBLEMS

The following are some of the problems that have resulted from human activity:

POLLUTION

The rise of industry in the past two centuries has led to greatly increased air and water pollution. Exhaust from cars and factories, and liquid and solid wastes from manufacturing and urban centers cloud the air and clog water supplies. Oil spills cover parts of the ocean and shoreline. Since almost all living organisms depend upon clean air and water, pollution poses a severe threat to the survival of life on Earth. In addition to this general threat, pollution poses several special dangers:

Global Warming. The burning of fossil fuels like coal, oil, and natural gas has significantly increased amounts of carbon dioxide in the atmosphere. Carbon dioxide and water act together to wrap the planet in a "blanket," holding in heat. With increased amounts of carbon dioxide, less heat is able to escape, leading to a **"greenhouse effect."** As a result, average temperatures on Earth's surface are gradually rising.

The Ozone Layer. Free oxygen combines with oxygen molecules to create **ozone** (O_3) in the Earth's upper atmosphere. This ozone absorbs much of the sun's ultraviolet radiation. Without this ozone layer, ultraviolet radiation would cause mutations in the DNA of most living cells in land areas. The use of chlorofluorocarbons, or **CFCs**, as coolants in refrigerators and air conditioners threatens the ozone layer. Each CFC molecule can break down thousands of ozone molecules. As a result, an ozone "hole" has appeared in the Earth's atmosphere, leading to increased incidents of skin cancer. Countries have agreed to ban CFCs, although some still use them.

Pesticides. Poisonous chemicals are used to control insects that threaten crops, but these pesticides then become part of the water and soil, endangering other organisms as well, such as birds. These pesticides may also be absorbed in the crops we grow for food.

Acid Rain. When coal and oil are burned, their smoke pollutes the atmosphere. Many pollutants from factories and automobile exhausts turn into acids. These acids get washed out of the air when it rains. These pollutants are highly toxic, killing fish, destroying forests, eroding soil and further endangering the environment.

LOSS OF NATURAL RESOURCES

Some resources, like forests and underground water, can renew themselves after a period of time. Other resources, like oil and coal, are **nonrenewable**, and can only be used once. Many human activities, like burning fossil fuels, are using up Earth's nonrenewable resources, while other activities are using renewable resources at a faster rate than they can renew themselves. This threatens the ecosystems these resources are taken from.

DESTRUCTION OF NATURAL HABITATS

One of the greatest threats to the environment is the destruction of natural habitats. As the human population expands, more forests, wetlands, and grasslands are destroyed to build farms, factories, and cities. The destruction of tropical rain forests is one of the most dramatic examples of the loss of natural habitats. Tropical rain forests have the greatest **biodiversity** (*diversity of species*) and the greatest concentration of producers (*plant life*). The destruction of areas like the Amazon rain forest reduces the amount of oxygen in the atmosphere and leads to the extinction of many species. More plant and animal species now become extinct each year than at any other time since the extinction of dinosaurs. This is important, since some genetic material in species becoming extinct may hold the answer to curing many diseases.

Destruction in the Amazon rain forest

ENVIRONMENTAL SOLUTIONS

Environmentalists are struggling to find solutions to the rising threats to the environment. Some of the solutions that have been proposed include:

Lifestyle Changes: Reduce, Reuse, Recycle. To combat pollution, environmentalists advocate less garbage, less paper and plastic wrapping, and more recycling. They also favor the greater use of *biodegradable* materials.

Alternative Sources of Energy. Instead of burning fossil fuels, environmentalists propose the use of **alternative energy sources**, such as wind power, solar power, and hydrogen-fueled engines. This would reduce air and water pollution and global warming.

Conservation. To preserve biodiversity, environmentalists have introduced laws protecting *endangered species* and natural habitats. To conserve resources, they recommend the use of renewable resources. For example, agriculture can be sustainable if it makes use of renewable resources and works along with natural ecosystem processes, instead of against them.

WHAT YOU SHOULD KNOW

- ★ You should know that an ecosystem is an **interdependent** system made up of living organisms (**biotic components**) and their environment (**abiotic components**).
- ★ You should know that different species interact through predation, parasitism, competition, mutualism, and commensalism.
- ★ You should know how energy and nutrients are recycled through an ecosystem. Drastic events, such as a fire or climatic change, may trigger changes to an ecosystem.
- ★ You should know that human activities now threaten many of the world's ecosystems.

CHAPTER STUDY CARDS

Ecosystems

- ★ **Biotic Components.** Living organisms in an ecosystem.
- ★ **Abiotic Components.** The non-living components of an ecosystem, such as temperature, sunlight, water, minerals.
- ★ **Terrestrial Ecosystems.** Ecosystems on land, such as deserts, forests, grasslands.
- ★ **Aquatic Ecosystems.** Ecosystems in water, such as oceans, lakes, and rivers.
- ★ **Community.** Organisms of different species found in a single ecosystem.
- ★ **Population.** All of the organisms of the same species in a particular ecosystem.

Interaction of Biotic Components

- ★ **Predation.** When one organism (**predator**) kills another organism (**prey**) for food.
- ★ **Parasitism.** When a parasite feeds on another organism, known as the **host**.
- ★ **Competition.** When two or more species in an ecosystem share similar traits and require the same resources.
- ★ **Mutualism.** When a cooperative relationship exists in which two or more species mutually benefit.
- ★ **Commensalism.** Occurs in a relationship where one species benefits without doing any harm to the other species.

The Flow of Energy and Nutrients through an Ecosystem

- ★ **Autotrophs.** Producers that make their own food. Autotrophs capture the sun's energy, turning it into carbohydrates through photosynthesis.
- ★ **Heterotrophs.** Organisms unable to make their own food that need to consume others or their wastes for energy.
  - **Herbivores.** Eat plants or phytoplankton.
  - **Carnivores.** Eat other animals.
  - **Omnivores.** Eat both plants and animals.
  - **Decomposers.** Break down organisms and wastes into organic molecules.

Impact of Humans on Ecosystems

- ★ **Environmental Problems:**
  - **Global Warming.** Increased CO_2 from burning fossil fuels is raising temperatures.
  - **Ozone Layer.** Ozone layer, which absorbs much of sun's ultraviolet radiation, is being destroyed by CFCs.
  - **Pesticides.** Pesticides can poison water, soil and the food we eat.
  - **Acid Rain.** Air pollutants turn into acids that are highly toxic.
- ★ **Loss of Natural Resources.**
- ★ **Destruction of Natural Habitats.**
  - Threatens loss of **Biodiversity**.

CHECKING YOUR UNDERSTANDING

Use the illustrations below to answer the following question:

1 Which of the illustrations to the right represents an ecosystem?

A. Illustration #1
B. Illustration #2
C. Illustration #3
D. Illustration #4

LS: F
10–16

This question examines your understanding of an ecosystem. Recall that an ecosystem consists of all of the living organisms in an area, together with their nonliving environment. Although Illustration #1 shows different species, they do not represent an ecosystem. Only Illustration #4 has both the biotic and abiotic components necessary for an ecosystem.

Now try answering some additional questions on your own about various types of ecosystems and how human activities can affect ecosystems.

2 Which of the following can be described as a population in an ecosystem?

A. all the honey bees in an orchard
B. all the plants and animals in a pond
C. all biotic and abiotic factors in a pond
D. all living things in Earth's atmosphere

- Examine the Question
- Recall What You Know
- Apply What You Know

LS: F
10–16

3 Which action by humans has had the most beneficial impact on the environment?

A. use of pesticides to regulate insect populations
B. importing of Japanese beetles into the United States
C. over-hunting of predators to prevent the death of prey animals
D. reforestation of trees and plants

LS: G
10–18

Use the passage below to answer questions 4–8:

Snowshoe Hares

The snowshoe hare (*Lepus americanus*) is the prey of many woodland animals. Foxes, coyotes, and weasels hunt the hares by day, while the great horned and barred owls hunt them at night. The snowshoe hare survives this predation partly through the effect of changing environmental conditions on its expression of genes for fur color, creating an effective camouflage. In winter months, the hare has white fur, which blends into the snow. During the warmer months, its fur changes to a reddish brown.

Some favorite foods of the hare during the winter months are twigs of maple, birch, and apple trees. Grasses and clover replace this diet during the spring and summer. Hares tend to feed during the hours of dusk and dawn, when the light is low and its predators are inactive. The hares prefer habitats along streams, wetlands, and spruce forests. Their population tends to vary throughout the year because individual animals have a short lifespan. There seem to be cycles of about 10 years, during which the population varies from about 0.1 hare per acre up to 5 hares per acre. Some biologists suggest that the cycles are related to predation and depletion of the hares' food sources.

4 If food supply were the only limiting factor on the growth of the snowshoe hare population, which graph would best represent the likely relationship between the population of snowshoe hares and their food supply over a ten-year period?

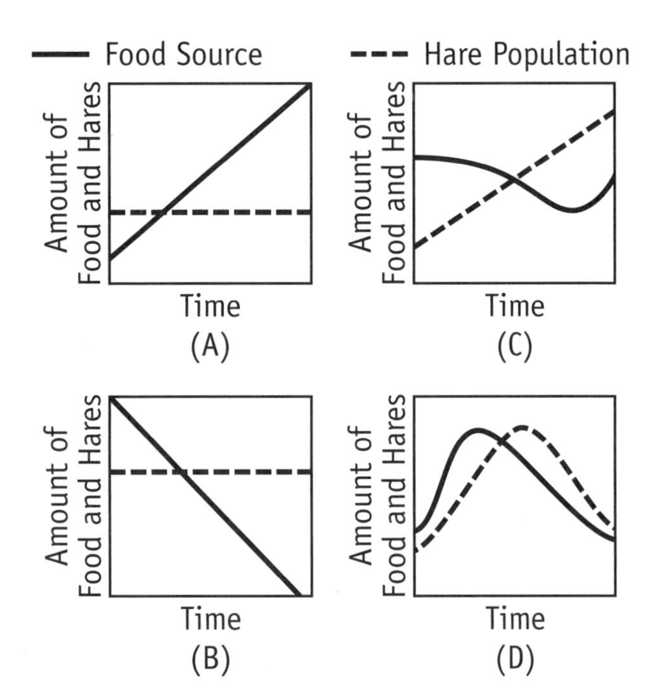

A. graph A
B. graph B
C. graph C
D. graph D

LS: F 10-16

5 If the population of snowshoe hares is reduced by disease, which change will most likely occur in the food web?

A. The fox and coyote population will increase.
B. The horned owl population will increase.
C. Grasses and clover will grow taller.
D. Apple trees will no longer grow.

LS: F 10-17

- Examine the Question
- Recall What You Know
- Apply What You Know

6 What is the genus of the snowshoe hare?

A. animal
B. *Lepus*
C. *americanus*
D. snowshoe

LS: E
10–12

7 Based on its diet, the snowshoe hare would be classified as

A. an herbivore
B. a carnivore
C. a decomposer
D. an autotroph

LS: D
10–9

8 Since foxes, coyotes, and weasels hunt snowshoe hares, they would be classified as

A. abiotic
B. predators
C. parasites
D. producers

LS: F
10–15

---

9 Arcaia trees in a forest ecosystem provide food for a species of ants that lives on them. The ants defend the trees from grasshoppers and beetles. This relationship between the ants and acacia trees is best described as

A. commensalism
B. mutualism
C. parasitism
D. competition

LS: F
10–15

- Examine the Question
- Recall What You Know
- Apply What You Know

10 Extended-Response Question

The food web diagram below shows the links that exist between various organisms in a particular ecosystem. Identify two different relationships in this food web. Describe the flow of energy among the organisms for each relationship you select. (*4 points*).

LS: D
10–9

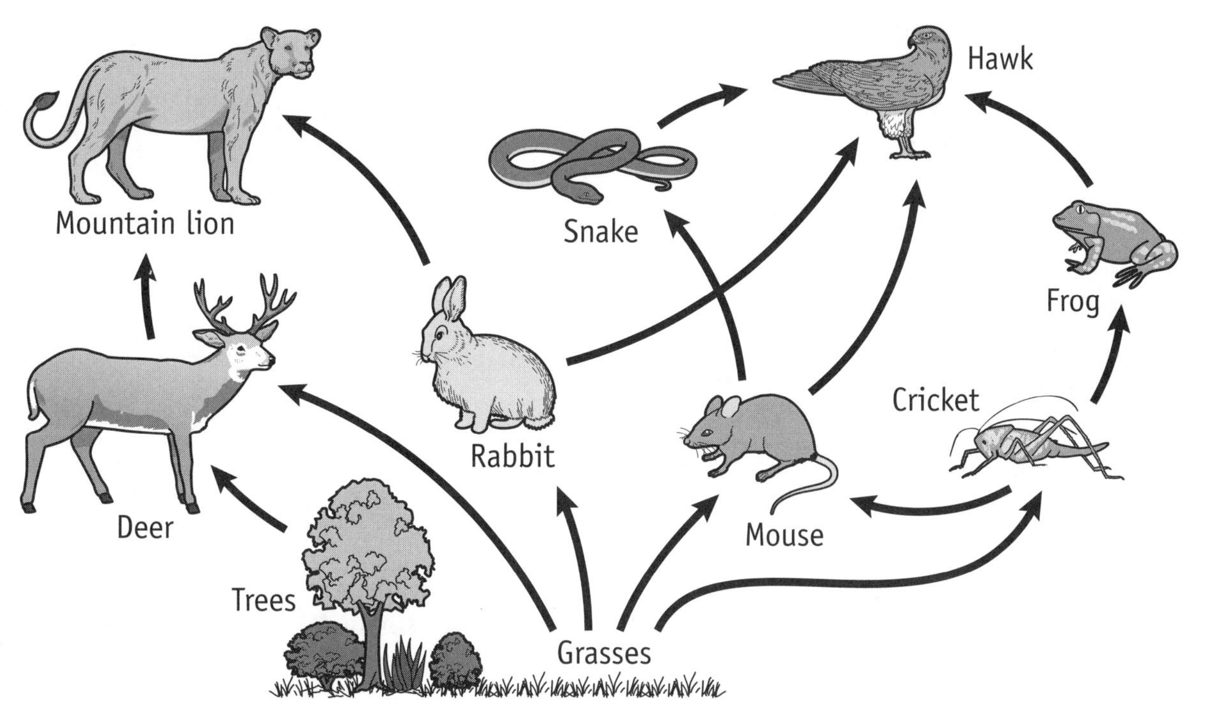

CHECKLIST OF BENCHMARKS IN THE LIFE SCIENCES UNIT

Directions. Now that you have completed this unit, place a check mark (✔) next to those benchmarks you understand. If you have trouble recalling information connected with one of the benchmarks, review the chapter indicated in the brackets for the items you do not recall.

- ❑ You should be able to explain that cells are the basic unit of structure and function of living organisms, that once life originated all cells come from pre-existing cells, and that there are a variety of cell types. **[Chapter 15]**
- ❑ You should be able to explain the characteristics of life as indicated by cellular processes and describe the process of cell division and development. **[Chapter 15]**
- ❑ You should be able to explain the genetic mechanisms and molecular basis of inheritance. **[Chapter 16]**
- ❑ You should be able to explain the flow of energy and the cycling of matter through biological and ecological systems (cellular, organismal and ecological). **[Chapters 15 & 18]**
- ❑ You should be able to explain how evolutionary relationships contribute to an understanding of the unity and diversity of life. **[Chapter 17]**
- ❑ You should be able to explain the structure and function of ecosystems and relate how ecosystems change over time. **[Chapter 18]**
- ❑ You should be able to describe how human activities can impact the status of natural systems. **[Chapter 18]**
- ❑ You should be able to describe a foundation of biological evolution as the change in gene frequency of a population over time. You should be able to explain the historical and current scientific developments, mechanisms and processes of biological evolution, and to describe how scientists continue to investigate and critically analyze aspects of evolutionary theory. **[Chapter 17]**
- ❑ You should be able to explain how natural selection and other evolutionary mechanisms account for the unity and diversity of past and present life forms. **[Chapter 17]**
- ❑ You should be able to summarize the historical development of scientific theories and ideas, and describe emerging issues in the study of life sciences. **[Chapters 15–18]**

A PRACTICE OGT IN SCIENCE

This chapter consists of a complete practice **OGT in Science**. Before you begin, let's review a few directions for the test:

★ **Answer All Questions.** The OGT in Science consists of multiple-choice, short-answer, and extended-response questions. Do not leave questions unanswered. There is no penalty for guessing. Blank answers are counted as wrong.

★ **Use the "E-R-A" Approach.** Remember to carefully examine the question to understand what the question is asking. Next, recall what you have learned about that particular topic in science. Finally, apply your knowledge to answer the question.

★ **Use the Process of Elimination.** When answering a multiple-choice question, it should be clear that certain choices are wrong. They will be irrelevant, lack a connection to the question, or be inaccurate. After you eliminate incorrect choices, select the best response that remains. Remember that often your first guess is correct.

★ **Revisit Difficult Questions.** It is possible that you will come across several difficult questions. If you run into a difficult question, do not be discouraged. Circle the question, or put a mark (✔) next to it. Answer it as best you can and move on to the next question. At the end of the test, go back and reread any questions you marked. Sometimes the answer to a difficult question might become clearer to you with a second reading.

★ **For Short and Extended-Response Questions.** Read the directions carefully. If time allows, make an answer box or notes before answering a short-answer or extended-response question. This will allow your responses to be more organized. Address all parts of the question. If you have time left at the end, check your work and correct any errors.

★ **When You Finish.** When you finish the test, review your work and make sure you have answered all the questions. Do not disturb other students.

As with every question in this book, this practice **OGT in Science** indicates the standard/benchmark and grade level indicator tested by each question. This information is provided to help you and your teacher identify any areas you might still need to study.

Good luck!

CHAPTER 19

A PRACTICE OGT IN SCIENCE

Directions: Each question is followed by four choices. Read each question carefully. Then circle or write down the letter of the choice that is the correct answer.

1 **Students in a laboratory have observed their classmates switching the labels on the liquids used in the lab. What is the most important reason the students should report this to their teacher?**

SI: A 9–2

A. It will affect the experiment if the wrong liquid is used.
B. Using the wrong liquid may cause someone to be injured.
C. Rules that are broken should be reported.
D. Teachers are the only people allowed to use the liquid.

Use the information in the passage to answer the following question.

POLYWATER

In 1966, a Russian scientist named N. N. Fedyakin claimed to have discovered a new form of water, polywater. Formed by heating water and letting it condense in narrow quartz tubes, polywater had a boiling point higher than 100° Celsius and a freezing point lower than 0°C.

Over the next several years, Fedyakin published dozens of papers in the scientific literature claiming to describe the properties of polywater. Because polywater could only be formed in very small amounts, none was ever available for analysis by others. When samples were finally analyzed by other scientists, polywater proved to have a variety of other substances in it. Using high powered electron microscopes, scientists revealed that polywater actually consisted of fine pieces of matter suspended in ordinary water. Following this revelation, all claims of the existence of polywater disappeared.

2 **The "discovery" of polywater by N.N. Fedyakin raised serious questions among some scientists. Explain two reasons why other scientists should have been concerned about the possible existence of polywater. (*2 points*).**

SK: A 9–4

____

Use the chart below showing the classification of four animals to answer the following question.

	Animals			
	1	2	3	4
Phylum	Chordata	Chordata	Chordata	Chordata
Class	Mammalia	Mammalia	Mammalia	Mammalia
Order	Lagomorpha	Lagomorpha	Lagomorpha	Lagomorpha
Family	Leporidae	Leporidae	Leporidae	Leporidae
Genus	Lepus	Sylvilagus	Pedetes	Lepus
Species	articus	floridanus	capensis	townsendii

3 Which two animals are the most closely related?

LS: I
10–12

A. Animals 1 and 2
B. Animals 1 and 4
C. Animals 2 and 3
D. Animals 3 and 4

---

4 Different isotopes of the same element have different

PS: A
9–1

A. atomic numbers.
B. numbers of protons.
C. numbers of neutrons.
D. numbers of electrons.

5 Which phrase best describes cellular respiration, a process that occurs continuously in the mitochondria of eukaryotic cells?

LS: B
10–3

A. removal of oxygen from the cells of an organism
B. conversion of light energy into the chemical bond energy of organic molecules
C. transport of materials within cells and throughout the bodies of multicellular organisms
D. changing of stored chemical energy to a form of energy usable by cells.

6 A space shuttle is traveling in frictionless outer space with a limited quantity of fuel. What action could the astronauts aboard the shuttle take to increase their rate of acceleration?

PS: D
9–23

A. reduce their velocity
B. eject cargo to reduce their mass
C. change their direction
D. turn off their rocket engines in order to glide

7 Viruses present an exception to cell theory, but they share some important characteristics with living things. What is one of these characteristics?

LS: A 10–1

A. They are made up of specialized cells.
B. They contain genetic material like DNA.
C. They cannot reproduce without a host.
D. They contain chlorophyll.

8 In the diagram below, the dark dots indicate small molecules. As shown by the arrows, these molecules are moving out of the cells. The number of dots inside and outside of the two cells represents the relative concentrations of the molecules inside and outside of the cells.

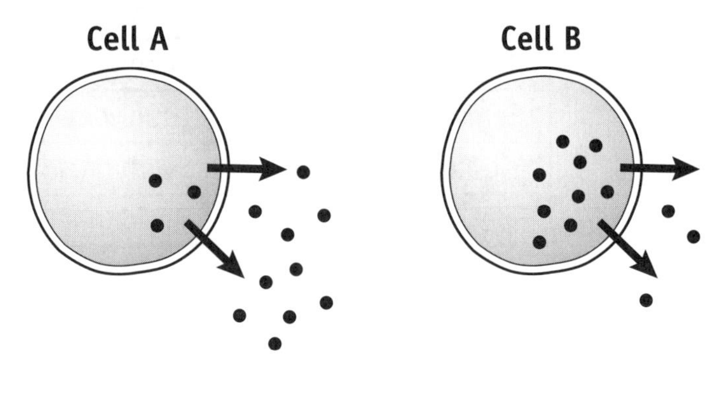

ATP is needed to move the molecules out of the cell by

LS: B 10–3

A. cell *A*, only.
B. cell *B*, only.
C. both cell *A* and cell *B*.
D. neither cell *A* nor cell *B*.

9 In a certain area of undisturbed layers of rock, fossils of horseshoe crabs may be found in the upper layer, and a lower layer contains fossils of trilobites, extinct aquatic arthropods that resemble modern horseshoe crabs in structure and appearance. This information suggests that

LS: E 10–24

A. horseshoe crabs will soon become extinct just as trilobites did.
B. horseshoe crabs and trilobites are unrelated organisms.
C. horseshoe crabs may have evolved from trilobites through natural selection.
D. trilobites may have evolved from horseshoe crabs through genetic drift.

10 Which organisms are correctly matched with their processes for releasing and acquiring energy?

LS: D 10–10

A. Maple trees – fermentation, German Shepard – cellular respiration, Baker's yeast– photosynthesis
B. Maple trees – fermentation, German Shepard – cellular respiration, Archaebacteria– photosynthesis
C. Maple trees – cellular respiration, Baker's yeast– fermentation Archaebacteria– chemosynthesis
D. German Shepard – cellular respiration, Baker's yeast– fermentation Archaebacteria– chemosynthesis

Use the following information to answer questions 11 to 14

Students place 10g of baking soda (sodium bicarbonate) into a beaker containing 100 mL of water, and stir until the baking soda is completely dissolved. Students place 10g of lemon juice into a second beaker containing 100 mL of water, and stir until the juice is thoroughly mixed. Using litmus paper, they record the pH value of each solution and take the temperature of each solution. They combine the two solutions and the mixture is gently stirred. The students record the temperature of the mixed solution five minutes after the liquids are combined and sixty minutes later. They also use litmus paper to determine pH value. The data from the experiment is recorded below and on the bar graphs as shown.

Baking soda solution:
Temperature: 23°C
pH Value: 10

Lemon juice solution:
Temperature: 23°C
pH Value: 4

Combined solution:
Temperature after 5 minutes: 19°C
Temperature after 60 minutes: 23°C
pH Value: 8

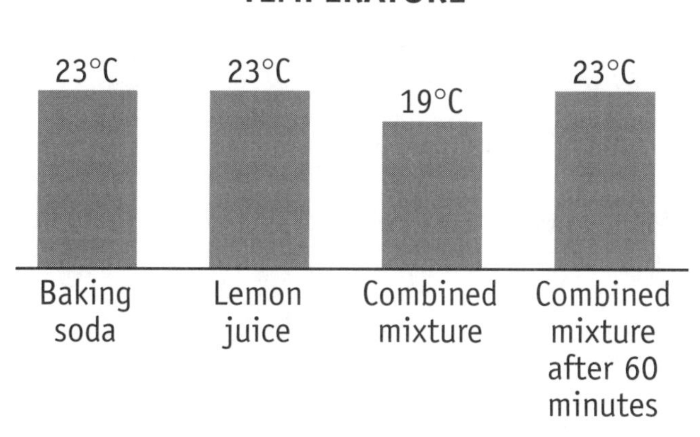

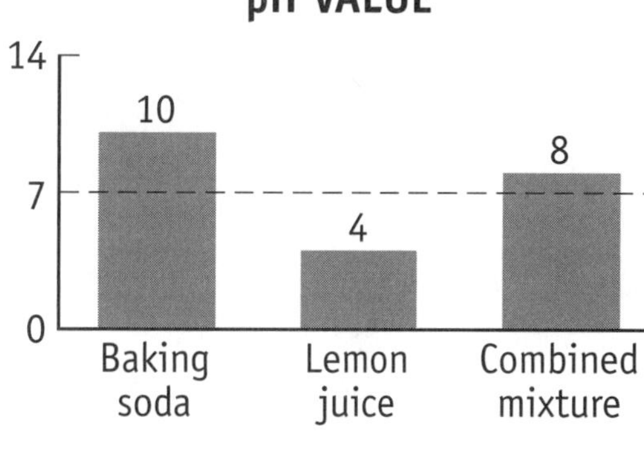

11 Based on the pH values recorded by the students, the combined solution is

A. a basic salt.
B. a strong base.
C. neutral.
D. an acidic salt.

PS: C 9–8

12 Based on the data, what happened when the two solutions were mixed together?

A. The reactants absorbed energy while combining.
B. Energy was released by a chemical reaction.
C. There was no change in the kinetic energy of the particles in the solution.
D. Combustion occurred when one of the reactants combined with oxygen.

PS: F 9–16

13 Which best describes the solution of baking soda and water?

A. an element
B. a compound
C. a homogeneous mixture
D. a heterogeneous mixture

PS: C 9–9

14 **Identify two dependent variables in this experiment. Then describe one way that the students might communicate their results and explain why communicating experimental results is an important part of any experiment. (*4 points*).**

SI: A 9–3 | SK:B 9–2

---

15 **A group of students found a large number of frogs missing some of their legs in a local pond. The students discussed what they saw and came up with a possible explanation: the frogs suffered from genetic mutations caused by the use of pesticides by local farmers. The students' explanation of what happened to the frogs provides an example of**

A. an observation.

B. a set of data.

C. a theory.

D. an inquiry.

SK: A 9–5

Use the diagram below to answer the following question.

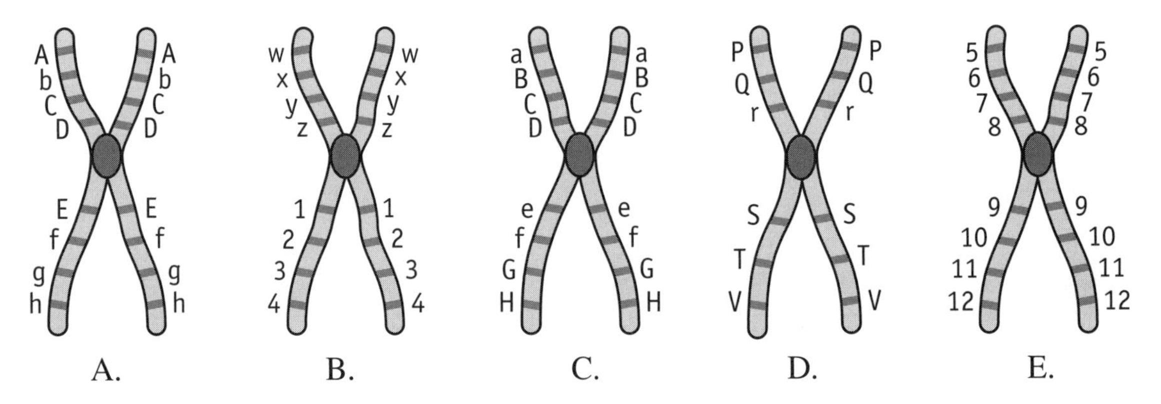

16 **Which of the chromosomes represented above form a homologous pair?**

A. A and B

B. B and C

C. C and A

D. D and E

LS: B 10–4

17 **State one example of a predator-prey relationship found in the food web. Indicate which organism is the predator and which is the prey. (*2 points*).**

LS: D 10–15

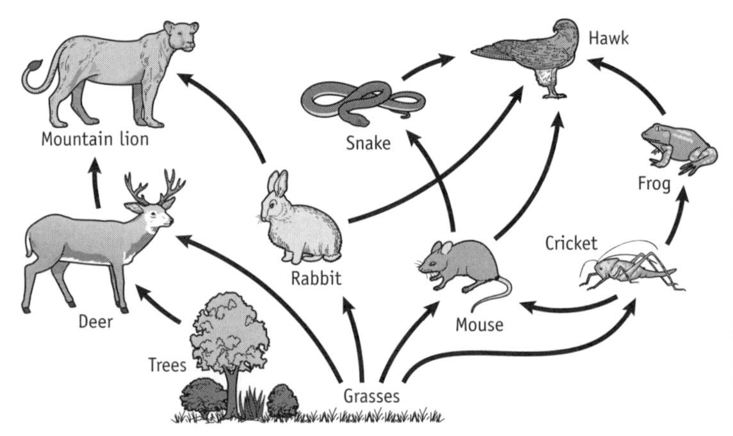

Use the pedigree chart below to answer questions 18 and 19.

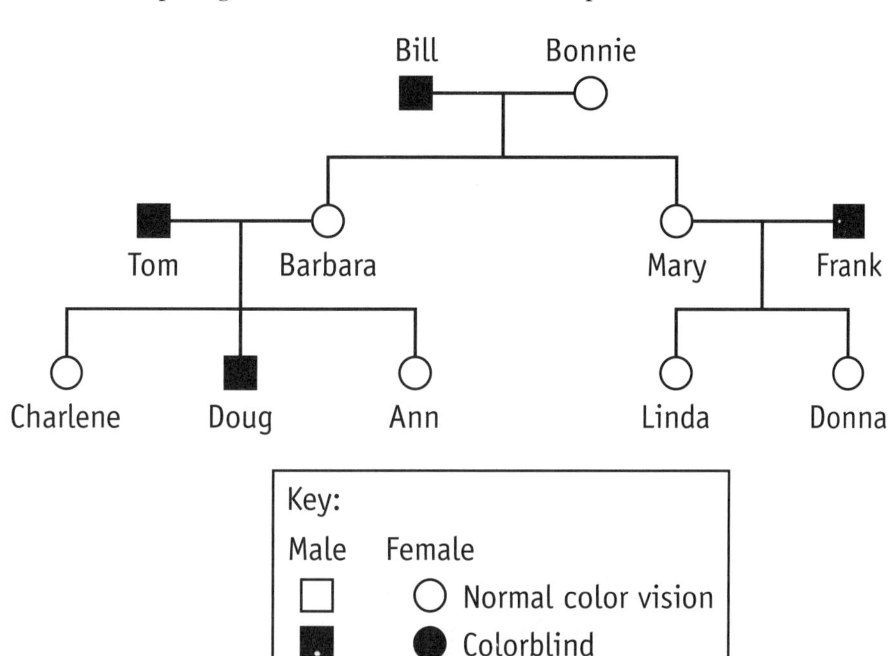

18 **The pedigree chart above represents the inheritance of color blindness through three generations. Color blindness is a sex-linked, recessive trait, carried on the X chromosome. A woman must inherit the trait from both parents to be colorblind. Barbara is expecting another child. What is the probability that Barbara and Tom's new baby will be colorblind?**

A. 0%
B. 25%
C. 50%
D. 100%

LS: C 10–8

19 **Which statement about the genotype of Linda and Donna regarding color blindness is correct?**

A. Both carry one recessive allele.
B. Linda is a carrier, and Donna is homozygous dominant.
C. Both are homozygous recessive.
D. Linda is homozygous dominant, and Donna is a carrier.

LS: C 10–8

---

20 **Which of the following identifies an important risk in using nuclear energy to power electricity plants?**

A. Americans would become more dependent on foreign oil.
B. Nuclear reactors produce radioactive waste that is difficult to dispose of.
C. The amount of oil reserves trapped below Earth's surface is limited.
D. Nuclear reactors create less air pollution than traditional fossil fuels.

ST: A 10–2

21 The diagram below shows a granite block being pushed at constant speed across a horizontal concrete floor by a force parallel to the floor.

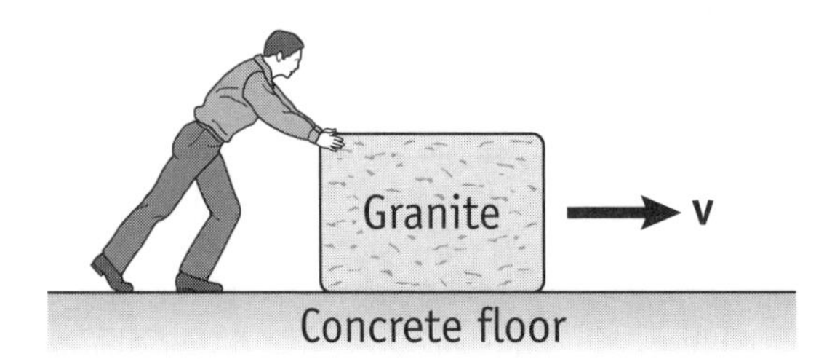

Under these conditions, which force prevents the block from accelerating?

A. friction
B. gravity
C. kinetic energy
D. inertia

PS: D 9–25

22 A student wishes to determine the volume of a rock. She fills a graduated cylinder with water up to 500 mL (1mL = 1 cm^3). Then she places the rock in the graduated cylinder. The diagram at the right shows what she observes. What is the volume of the rock?

A. 220 cm^3
B. 260 cm^3
C. 720 cm^3
D. 760 cm^3

SI: A 10–4

1000 mL
1000
900
800
700
600

23 Which graph best represents the relationship between the gravitational potential energy of an object and the object's height above the surface of Earth?

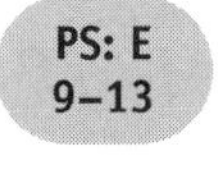

PS: E 9–13

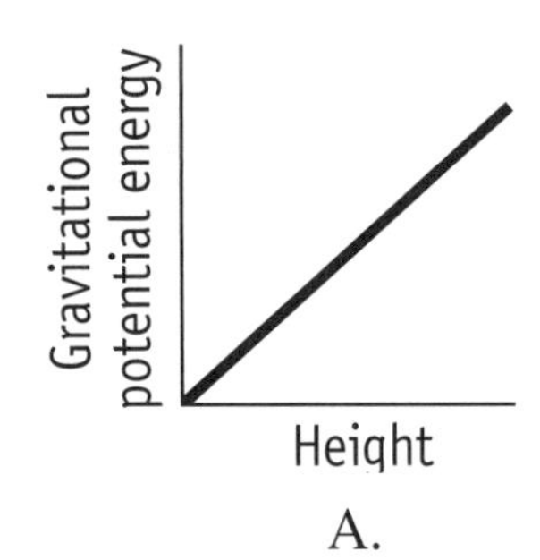

A.

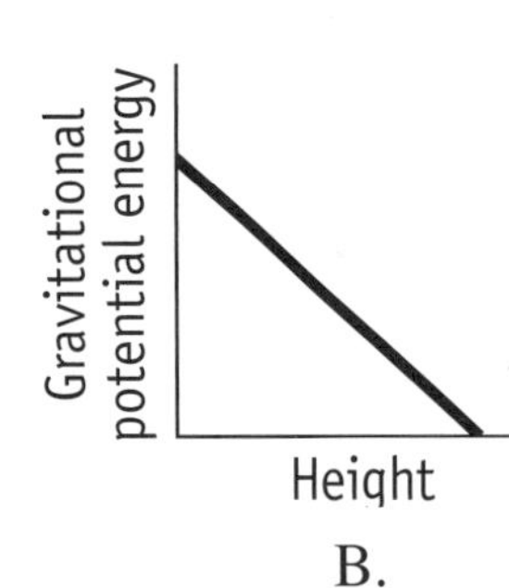

B.

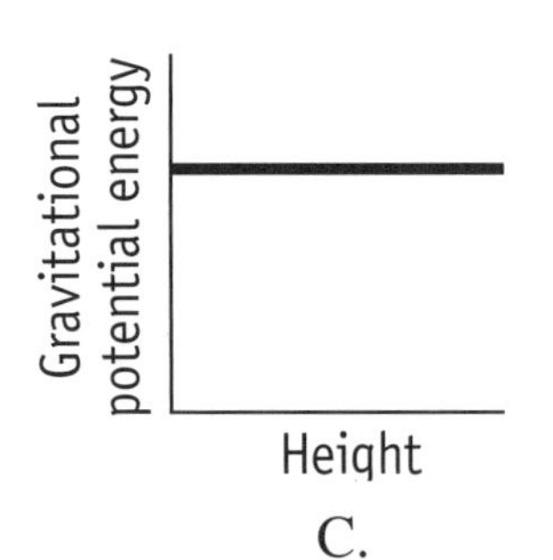

C.

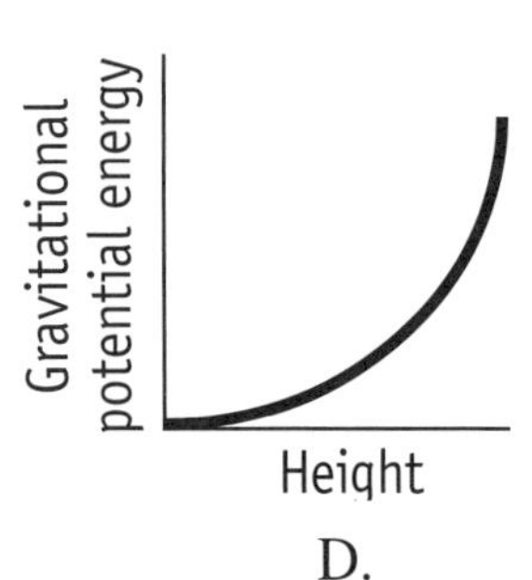

D.

24 Technology often provides people with new products and processes for making their lives better and easier. Sometimes, however, the adoption of a new technology brings new risks. Scientists often help evaluate what technologies should be adopted. Identify two factors that should be considered when deciding whether or not to adopt a new technological design. (*2 points*)

ST: A 10–3

25 The gravitational pull of the moon is the greatest influence on the water levels of Earth's ocean tides. If the distance between the moon and Earth were to *decrease* steadily for one week, which water-level changes would be expected to occur?

ES: A
9–3

A. High tides would get higher, and low tides would get lower.
B. High tides would get lower, and low tides would get higher.
C. Both high tides and low tides would get higher.
D. Both high tides and low tides would get lower.

26 The cross sections below show different patterns of air movement in the atmosphere above the ocean. Air temperatures on the ocean's surface are indicated in each cross section. Which cross section shows the most likely pattern of air movement in Earth's atmosphere that would result from the surface air temperatures shown?

ES: B
9–4

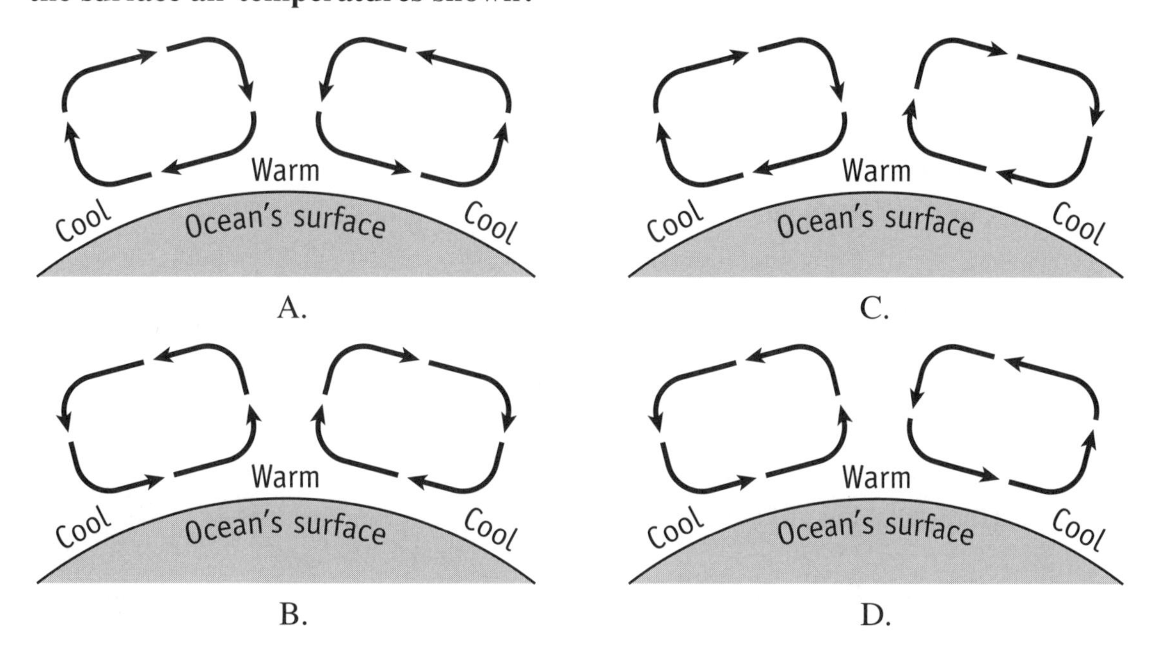

27 Each of the objects below has a different amount remaining of the same radioactive material X. Which object is the oldest?

ES: C
10–3

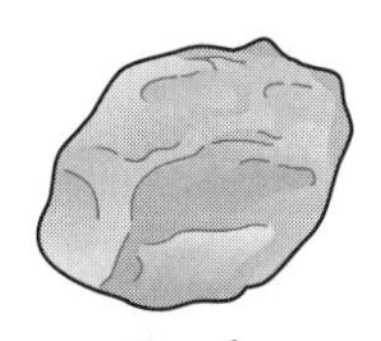

Rock
10% of the radioactive material remains
A.

Wood
33% of the radioactive material remains
B.

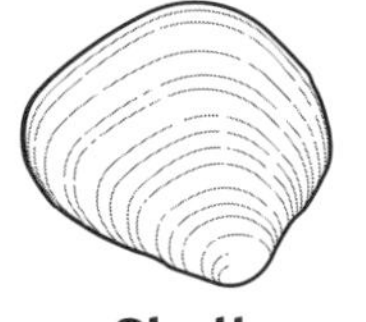

Shell
41% of the radioactive material remains
C.

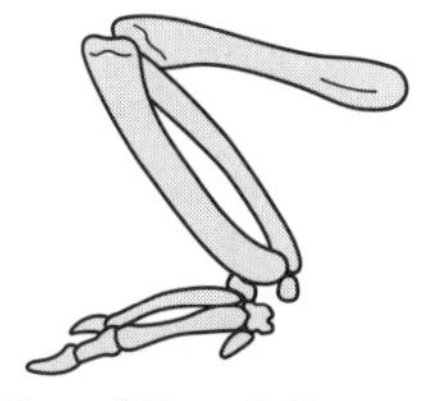

Fossilized Bone
52% of the radioactive material remains
D.

Below is the cross section of two tectonic plates and the boundary between them. The arrows indicate the direction of plate movement.

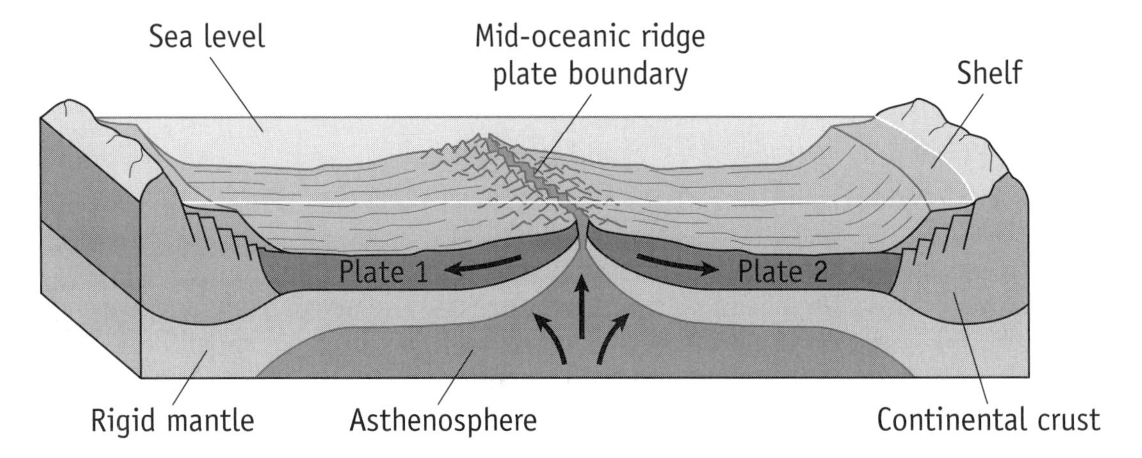

28 The mid-oceanic ridge of this cross section represents

ES: E 9–7

A. the convergence of two plates.
B. the divergence of two plates.
C. the subduction of one plate under another.
D. transform faulting between two plates.

29 Which geologic events occur most often at this mid-oceanic ridge plate boundary?

ES: E 9–6

A. magnetic pole reversals and cooling of ocean water
B. meteorite impacts and tilting of shore lines
C. hydrosphere pollution and adiabatic breathing
D. earthquakes and volcanic eruptions

---

30 What evidence led Alfred Wegener to conclude a century ago that all continents were once joined together and then drifted apart?

ES: F 9–8

A. Wegener discovered ancient written records, witnessing these events.
B. Fossils of the same animals were found in both South America and Africa.
C. Scientists in Wegener's time discovered magnetic reversals on the deep ocean floor.
D. Wegener found evidence of a tidal wave strong enough to move continents.

31 Which two factors are abiotic factors that affect organisms in marine biomes?

LS: F 10–15

A. amount of algae and wide temperature variations
B. amount of carbon dioxide and variety of producer organisms
C. amount of moisture and variety of consumer organisms
D. amount of oxygen and concentration of dissolved salts

Use the partial periodic table to answer questions 32 to 34

Key	
6	Atomic number
C	Symbol
Carbon	Name
12.0107	Average atomic mass

Period	IA 1	IIA 2	IIIA 13	IVA 14	VA 15	VI 16	VII 17	VIIIA 18
1	1 H Hydrogen 1.00794							2 He Helium 4.0026
2	3 Li Lithium 6.941	4 Be Beryllium 9.0122	5 B Boron 10.811	6 C Carbon 12.0107	7 N Nitrogen 14.0067	8 O Oxygen 15.9994	9 F Fluorine 18.9984	10 Ne Neon 20.1797
3	11 Na Sodium 1.00794	12 Mg Magnesium 24.3050	13 Al Aluminum 26.98154	14 Si Silicon 28.0855	15 P Phosphorus 30.9738	16 S Sulfur 32.065	17 Cl Chlorine 35.4527	18 Ar Argon 39.948
4	19 K Potassium 1.00794	20 Ca Calcium 40.078						

32 Which of the following groups on the Periodic Table is likely to form positively charged ions?

A. Group 1
B. Group 15
C. Group 17
D. Group 18

PS: B 9–5

33 Why are the elements in Group 18, known as the noble gases, unlikely to combine with other substances?

A. They have low boiling points.
B. They have unstable nuclei.
C. They have rigid, crystalline structures.
D. They have a completed outer energy level of electrons.

PS: B 9–7

34 Based on the Periodic Table of the Elements (above), how many protons and neutrons does a typical nitrogen (N) atom have? (*2 points*)

PS: A 9–4

35 Environmental groups have become increasingly alarmed at the fact that many people treat our planet as though it were something that could be tossed into a trash can. Describe two problems that human activities are now causing to our planet's natural resources. For each problem, identify one way of reducing or solving that problem. (*4 points*)

ES: D
10–7

36 Why does white light divide into different colors when it passes through a triangular glass prism?

PS: G
9–20

A. The light waves interfere with each other, creating patterns.
B. Each color has a different wavelength, which refracts at a different angle.
C. Part of the white light is absorbed by the glass
D. White light consists of invisible particles traveling at different speeds.

37 Which of the following provides evidence in support of the Big Bang Theory of the universe?

ES: A
9–2

A. the existence of gravity
B. background cosmic radiation
C. the decay of radioactive isotopes
D. the heliocentric theory

38 Rocks are classified as igneous, sedimentary, or metamorphic based primarily on their

ES: E
9–6

A. texture.
B. crystal or grain size.
C. method of formation.
D. mineral composition.

GLOSSARY

A

Abiotic Conditions. Nonliving things or conditions, such as climate, which affect an ecosystem. [187]
Acceleration. The rate at which a velocity changes; often measured in m/s^2. [126]
Acid. A substance that forms a sour solution in water, and has a pH of less than 7. [118]
Active Transport. Any process in which a cell uses its energy to expel molecules. [153]
Adaptation. Adjustment to environmental conditions by an organism. [173]
Adaptive radiation. When a species evolves into various subgroups adapting differently to the environment. [175]
Alleles. Two genes that determine the same trait of an organism. [162]
Archaebacteria. Unicellular prokaryotes similar to first life on Earth. [173]
Atmosphere. The gaseous envelope surrounding Earth, made up mainly of nitrogen and oxygen. [82]
ATP. An organic compound used by living organisms to store and release energy. [154]
Atom. The smallest particle of an element that can exist either alone or in combination. [99]
Atomic Number. The number of protons in the nucleus of an atom. [100]

B

Base. A substance that forms a bitter solution in water, and has a pH of more than 7. [118]
Big Bang Theory. The theory that the universe began in a single explosion billions of years ago. [68]
Binary Fission. The process by which prokaryotic cells divide. [155]
Biodiversity. The degree of diversity of different species. [189]
Biome. Major type of ecological community (such as tropical rain forest, grassland, or desert). [83]
Biotechnology. Biological science combined with technology to meet human needs. [166]
Biotic Conditions. Conditions related to living things. [187]
Boiling Point. The temperature at which a liquid turns to gas. [117]

C

Cell. The smallest basic unit of structure and function of all living things. [149]
Cell Membrane. The membrane that holds the cells together and controls the entry of molecules and the interaction of cells with their environment. [151]
Cell Theory. The theory that all living things are made up of cells coming from pre-existing cells. [150]
Cell Wall. Cellulose wall around the cell membrane of a plant cell, providing structure and support. [152]
Cellular Respiration. Process by which cells break down carbohydrates to release energy. [154]
Chemical Change. A change in a substance resulting in a new substance. [111]
Chemical Property. Ability of a substance to react with other substances. [112]
Chemical Reaction. A process that involves rearrangement of the molecular or ionic structure of a substance, as opposed to a change in physical form. [104]
Chromosome. Threadlike structure in the cell nucleus carrying genetic information in DNA molecules. [155]
Classification. Systematic arrangement of items into groups such as genus and species. [176]
Climate. The average weather conditions of a place over a period of years. [83]
Commensalism. A relationship in which one organism benefits without harming the other. [184]
Community. All the different species of organisms living in a defined area. [182]

Compound. A substance formed from two or more elements chemically combined in fixed proportions: for example, water (H_2O) is a compound. **[104]**
Conduction. The process by which heat or electricity is transmitted through an object or between objects in direct contact. **[137]**
Conservation of Matter/Energy. Neither matter nor energy can be lost or destroyed. **[105]**
Consumer. A living organism that lives by eating other organisms; also known as a heterotroph. **[185]**
Control Group. A group used as a basis of comparison for checking the results of an experiment; the control group is compared to the experimental group. **[45]**
Convection. The circular motion that occurs in a heated fluid, as hot fluid rises and cooler fluid sinks. Convection is one way heat is transferred. Convection also occurs in Earth's mantle. **[75]**
Convergent Plate Boundaries. Areas where tectonic plates collide as they move toward each other. Convergent plate boundaries often lead to folding or subduction. **[76]**
Core. Center of Earth consisting mainly of extremely hot iron and nickel. **[74]**
Covalent Bond. Chemical bonds between atoms formed by the sharing of electrons. **[103]**
Crust. The thin, outer shell of Earth, made of rock. **[73]**
Cytoplasm. The main body of a cell, inside the cell membrane but outside the nucleus; consists of fluid (*cytosol*) and organelles. **[15]**

D

Data. Facts collected during an experiment or observed in nature. **[47]**
Divergent Plate Boundaries. Places where tectonic plates move apart from each other, creating a gap that is filled by magma; the magma then forms a ridge, such as the Mid-Atlantic Ridge. **[76]**
Decomposers. Organisms, such as bacteria and fungi, that break down dead organisms. **[185]**
Density. An object's mass divided by its volume (often measured in g/cm^3). **[119]**
Dependent Variable. A variable whose quantity changes as the result of a change made to the independent variable in an experiment. **[45]**
DNA. Deoxyribonucleic acid, a double helix of nucleotides which contains the genetic code and transmits traits in all living organisms. Watson and Crick developed the first DNA model. **[149]**
Dominant Trait. A trait, which if present in either allele, will appear in the individual. **[162]**

E

Ecosystem. An ecological system: a community of organisms and their environment. **[182]**
Electricity. A form of energy resulting from the flow of negatively charged particles (electrons). **[136]**
Electromagnetic Radiation. Waves of energy that can travel through a vacuum, including radio waves, infrared waves, visible light, ultraviolet light, gamma rays and x-rays. **[142]**
Electron. A subatomic particle with a negative electrical charge and almost no mass. **[99]**
Element. A fundamental substance with atoms of one kind. **[98]**
Endoplasmic Reticulum. Tubes that transport proteins and molecules in eukaryotic cells. **[151]**
Endothermic. Chemical reactions that require and absorb energy. **[138]**
Energy. The capacity for doing work; can be in such forms as nuclear, sound, thermal and light. **[133]**
Equilibrium. A state in which opposing forces or influences are balanced. **[184]**
Ethical Guidelines. Rules stating what scientists can and cannot do, such as when using animals. **[38]**
Eukaryotic Cells. Cells containing a membrane-covered nucleus and organelles. **[151]**
Evolution. Changes in the genetic composition of a population through successive generations. **[170]**
Exothermic. Chemical reactions that produce and release energy. **[138]**
Experiment. An attempt to test a hypothesis by gathering data under controlled conditions. **[33]**

F

Fermentation. Anaerobic process in which some cells, like yeast, obtain energy without oxygen. [155]
Fission. The splitting of an atomic nucleus, releasing large amounts of energy. [136]
Folding. Bending of Earth's crust at convergent plate boundaries, creating mountains. [76]
Food Chain or Food Web. An arrangement of the organisms of an ecosystem based on how they use each other as food sources; shows producers and different types of consumers. [186]
Force. An influence, which if applied to a moving body, results in an acceleration of that body. [133]
Fossil. Impression or traces of an organism of past geologic ages preserved in Earth's crust. [91]
Friction. The force from rubbing that resists the continued motion of two objects in contact. [129]
Fusion. The joining of atomic nuclei to form heavier nuclei, releasing enormous quantities of energy. [136]

G

Gas. A state of matter without precise shape or volume, in which particles are spread apart. [117]
Gene. The part of a chromosome that determines a specific trait. [162]
Genetic Drift. The process by which gene frequencies in small populations are changed. [175]
Genotype. The genetic make-up of an individual, showing both alleles governing a trait. [162]
Golgi Complex. Membranes that package and secrete molecules from a eukaryotic cell. [152]
Gravitational Potential Energy. Energy stored in an object based on its weight and height. [135]
Gravity. The attraction between any two objects, proportional to their masses, divided by the distance between them squared. [66]

H

Heliocentric Theory. The theory introduced by Nicolaus Copernicus that Earth circles the sun. [65]
Heredity. The sum of characteristics inherited from one's ancestors, based on genes. [161]
Heterozygous. Having two different genes for a particular trait, one recessive and one dominant. [162]
Homeostasis. The processes by which a living thing maintains a state of equilibrium. [153]
Homologous Chromosomes. A pair of chromosomes that govern the same traits. [157]
Homozygous. Having two similar genes for a particular trait (*e.g. both alleles are dominant*). [162]
Hydrosphere. All water on Earth's surface and atmosphere; mainly found in the oceans. [79]
Hypothesis. An educated guess that attempts to answer a scientific question. [44]

I, J, K

Igneous Rock. Rock made from cooled magma. [79]
Independent Variable. A variable that a scientist changes to find out how this change affects other variables in the experiment. [45]
Ion. An atom that carries a positive or negative electric charge from having lost or gained electrons. [102]
Isotope. Different atoms of the same element with the same atomic number but with different numbers of neutrons and a different atomic mass; for example, carbon-12 and carbon-14. [100]
Kinetic Energy. The energy of motion: $KE = 1/2\ mv^2$ [134]

L, M

Lithosphere. Earth's crust and upper mantle, about 80 km thick, made of shifting plates. [74]
Magma. Molten rock material within Earth; becomes lava at the surface. [76]
Mantle. The part of the interior of Earth that lies beneath the crust and above the central core; made of hot, semi-solid rock. [74]

Mass. The amount of matter an object has; it is proportional to its weight. **[98]**
Matter. Anything that occupies space and has mass. **[116]**
Meiosis. The process in which sex cells divide in half, allowing for genetic variation. **[156]**
Metamorphic Rock. Rock, such as marble or slate, made from igneous or sedimentary rock that has been changed by heat and pressure under Earth's surface. **[79]**
Metal. Group of elements, to the left of the Periodic Table, which lose electrons when ionized; metals are generally hard, shiny, ductile, and good conductors of heat and electricity. **[114]**
Mitochondria. Membrane-covered cell structure responsible for cellular respiration in eukaryotic cells. **[152]**
Mitosis. The process by which a cell's nucleus copies its chromosomes; these chromatids separate; and the nucleus divides in two to create two cells that are the same as the original cell. **[155]**
Mixture. Two or more substances mixed together but not chemically combined. **[111]**
Molecule. A group of atoms joined by covalent bonding. **[103]**
Mutation. A hereditary change in a gene caused randomly or by environmental conditions. **[166]**
Mutualism. Relationship between organisms of different species in which each benefits. **[184]**

N

Natural Selection. The theory that individuals having characteristics that aid in survival will produce more offspring, and that the proportion of such individuals in a species will gradually increase. **[171]**
Neutron. An uncharged particle found in the nucleus with a mass equal to a proton. **[99]**
Newton's Laws of Motion. (1) An object at rest stays at rest and an object in motion stays in motion unless an unbalanced force acts upon it; (2) Force equals mass times acceleration ($F = m \cdot a$); (3) for every action, there is an equal and opposite reaction. **[126]**
Nuclear Reaction. Reaction based on splitting or fusing of atomic nuclei, releasing vast energy. **[136]**
Nucleus. (1) The positively charged central portion of an atom that comprises nearly all of the atomic mass and that consists of protons and neutrons. (2) The part of the eukaryotic cell that is surrounded by a nuclear membrane and stores the cell's DNA. **[151]**

O, P, Q

Organelle. A specialized structure in a cell, often covered with a membrane. **[151]**
Parasite. An organism living in, with or on another organism in which a parasite obtains benefits from a host that it usually injures. For example, a tapeworm is a parasite. **[183]**
Periodic Table. A table showing all the known elements based on their atomic number, arranged into periods and groups based on their common properties and electron arrangement. **[100]**
pH Scale. A measure of the acidity or alkalinity of a chemical solution based on a scale of 0-14. **[118]**
Photosynthesis. The process by which green plants use light energy to convert carbon dioxide and water into carbohydrates, storing energy and releasing oxygen as a byproduct. **[154]**
Physical Properties. A property of a substance that may change without changing its chemical makeup: color, hardness, odor, state (gas, liquid, solid), boiling and freezing points. **[116]**
Plate Tectonic Theory. The theory that Earth's lithosphere consists of giant shifting plates. **[75]**
Predator. An animal that lives by eating prey; "predation" refers to the predator/prey relationship. **[183]**
Producer. An organism that produces its own food, such as a green plant; also known as an autotroph. **[185]**
Prokaryotic Cells. The oldest cells on Earth. A prokaryotic cell has DNA but no nuclear membrane; prokaryotes are unicellular; they divide by binary fission. Bacteria are prokaryotic cells. **[150]**
Proton. A particle in the nucleus of an atom with a positive electric charge and an atomic mass of one. **[99]**

R

Radiation. Waves of energy that can travel through a vacuum. **[137]**
Radioactive Decay. When an unstable nucleus decays, emitting subatomic particles and radiation. **[101]**
Radiometric Dating. A method used to date rocks or fossils by measuring the decay of radioactive substances. **[92]**
Reactant. A substance that enters into a chemical reaction. **[104]**
Recessive Trait. A genetic trait that appears only if shared by both alleles; otherwise, it is masked by the dominant trait but may reappear in later generations. **[162]**
Reflection. The bouncing back of light, heat or sound after hitting a surface. **[141]**
Refraction. Bending of a light ray or energy wave passing from one medium (*such as air*) into another (*such as glass*) caused by changes in the speed of the wave. **[141]**

S

Scientific Investigation. Methods scientists use to investigate the natural world and conduct experiments by asking a question, forming a hypothesis, making a prediction, testing the hypothesis, gathering data, drawing conclusions, and communicating results. **[42]**
Scientific Knowledge. The accumulation of all scientific work accomplished up to the present, based on careful observation, experimentation, and the testing of theories. **[35]**
Sedimentary Rock. Rock created in layers by the accumulation and compression of sediment. **[79]**
Seismic Waves. Waves of energy sent by earthquakes; measured by seismographs. **[73]**
Solid. A state of matter with definite density and shape, in which molecules are closely packed. **[116]**
Solution. A homogeneous mixture in which one substance is evenly dispersed in another. **[111]**
Sound Waves. Mechanical radiant energy that is transmitted by longitudinal pressure waves in a material medium (*such as air*) and is the objective cause of hearing. **[140]**
Species. A group of organisms of similar structure, capable of mating and producing offspring. **[177]**
Speciation. The creation of a new species, for example when subgroups are isolated for long periods. **[175]**
Star. A giant object in space, like the sun, that produces energy through nuclear fusion. **[67]**
Subduction Zone. In plate tectonics, a convergent boundary where a dense oceanic plate slides under a lighter continental plate. **[77]**
Substance. A form of matter with a uniform chemical structure; either an element or a compound. **[110]**

T, U

Technology. The application of scientific knowledge to meet human needs. **[55]**
Temperature. A measure of the average kinetic motion of particles in an object. **[134]**
Theory. A possible explanation of observations and data, which can be tested and revised. **[33]**
Thermal Energy. Heat energy based on the motion of molecules and atoms in a substance. **[137]**
Tides. The rising and falling of the oceans caused mainly by the gravitational pull of the moon. **[80]**
Trait. An inherited characteristic. **[162]**

V, W, X, Y, Z

Variable. A quantity that can change. **[44]**
Velocity. The rate of change of an object's position over time (*often measured in meters/second*). **[125]**
Water Cycle. Water passes from vapor in the atmosphere through precipitation to land or water surfaces, and then back again into the atmosphere as a result of evaporation and transpiration. **[79]**
Wave. A vibration or disturbance carrying energy, such as seismic, sound, or light waves. **[140]**

OHIO'S ACADEMIC STANDARDS IN SCIENCE

The following pages list all of Ohio's science standards with their related benchmarks and grade level indicators. These science standards have been referred to throughout this book. In particular, every question has been identified by its specific benchmark and grade level indicator. These are identified by an abbreviation for the standard and particular benchmark (such as "ES:A" for *Earth and Space Sciences* and *benchmark A*), and by the grade and number of the indicator (*"9-3" for grade nine, grade level indicator 3.*)

ES: A
9–3

EARTH AND SPACE SCIENCES: ES

BENCHMARKS

ES:A Explain how evidence from stars and other celestial objects provide information about the processes that cause changes in the composition and scale of the physical universe.

ES:B Explain that many processes occur in patterns within the Earth's systems.

ES:C Explain the 4.5 billion-year-history of Earth and the 4 billion-year-history of life on Earth based on observable scientific evidence in the geologic record.

ES:D Describe the finite nature of Earth's resources and those human activities that can conserve or deplete Earth's resources.

ES:E Explain the processes that move and shape Earth's surface.

ES:F Summarize the historical development of scientific theories and ideas, and describe emerging issues in the study of Earth and space sciences.

GRADE NINE GRADE LEVEL INDICATORS

9–1. Describe that stars produce energy from nuclear reactions and that processes in stars have led to the formation of all elements beyond hydrogen and helium.

9–2. Describe the current scientific evidence that supports the theory of the explosive expansion of the universe, the Big Bang, over 10 billion years ago.

9–3. Explain that gravitational forces govern the characteristics and movement patterns of the planets, comets and asteroids in the solar system.

9–4. Explain the relationships of the oceans to the lithosphere and atmosphere (e.g., *transfer of energy, ocean currents and landforms*).

9-5. Explain how the slow movement of material within Earth results from:
a. thermal energy transfer (conduction and convection) from the deep interior;
b. the action of gravitational forces on regions of different density.

9–6. Explain the results of plate tectonic activity (e.g., *magma generation, igneous intrusion, metamorphism, volcanic action, earthquakes, faulting and folding*).

9–7. Explain sea-floor spreading and continental drift using scientific evidence (e.g., *fossil distributions, magnetic reversals and radiometric dating*).

9–8. Use historical examples to explain how new ideas are limited by the context in which they are conceived; are often initially rejected by the scientific establishment; sometimes spring from unexpected findings; and usually grow slowly through contributions from many different investigators (e.g., *heliocentric theory and plate tectonics theory*).

GRADE TEN GRADE LEVEL INDICATORS

10–1. Summarize the relationship between the climatic zone and the resultant biomes. (This includes explaining the nature of the rainfall and temperature of the mid-latitude climatic zone that supports the deciduous forest.)

10–2. Explain climate and weather patterns associated with certain geographic locations and features (e.g., *tornado alley, tropical hurricanes and lake effect snow*).

10–3. Explain how geologic time can be estimated by multiple methods (e.g., *rock sequences, fossil correlation and radiometric dating*).

10–4. Describe how organisms on Earth contributed to the dramatic change in oxygen content of Earth's early atmosphere.

10–5. Explain how the acquisition and use of resources, urban growth and waste disposal can accelerate natural change and impact the quality of life.

10–6. Describe ways that human activity can alter biogeochemical cycles (e.g., *carbon and nitrogen cycles*) as well as food webs and energy pyramids (e.g., *pest control, legume rotation crops vs. chemical fertilizers*).

10–7. Describe advances and issues in Earth and space science that have important long-lasting effects on science and society (e.g., *geologic time scales, global warming, depletion of resources and exponential population growth*).

LIFE SCIENCES: LS

BENCHMARKS

LS:A Explain that cells are the basic unit of structure and function of living organisms, that once life originated all cells come from pre-existing cells, and that there are a variety of cell types.

LS:B Explain the characteristics of life as indicated by cellular processes and describe the process of cell division and development.

LS:C Explain the genetic mechanisms and molecular basis of inheritance.

LS:D Explain the flow of energy and the cycling of matter through biological and ecological systems (*cellular, organismal and ecological*).

LS:E Explain how evolutionary relationships contribute to an understanding of the unity and diversity of life.

LS:F Explain the structure and function of ecosystems and relate how ecosystems change over time.

LS:G Describe how human activities can impact the status of natural systems.

LS:H Describe a foundation of biological evolution as the change in gene frequency of a population over time. Explain the historical and current scientific developments, mechanisms and processes of biological evolution. Describe how scientists continue to investigate and critically analyze aspects of evolutionary theory.

LS:I Explain how natural selection and other evolutionary mechanisms account for the unity and diversity of past and present life forms.

LS:J Summarize the historical development of scientific theories and ideas, and describe emerging issues in the study of life sciences.

GRADE NINE GRADE LEVEL INDICATORS

No Indicators present in this grade for Life Sciences standard.

GRADE TEN GRADE LEVEL INDICATORS

10–1. Explain that living cells
- **a.** are composed of a small number of key chemical elements (*carbon, hydrogen, oxygen, nitrogen, phosphorus and sulfur*)
- **b.** are the basic unit of structure and function of all living things
- **c.** come from pre-existing cells after life originated, and
- **d.** are different from viruses

10–2. Compare the structure, function and interrelatedness of cell organelles in eukaryotic cells (e.g., *nucleus, chromosome, mitochondria, cell membrane, cell wall, chloroplast, cilia, flagella*) and prokaryotic cells.

10–3. Explain the characteristics of life as indicated by cellular processes including
- **a.** homeostasis
- **b.** energy transfers and transformation
- **c.** transportation of molecules
- **d.** disposal of wastes
- **e.** synthesis of new molecules

10–4. Summarize the general processes of cell division and differentiation, and explain why specialized cells are useful to organisms and explain that complex multicellular organisms are formed as highly organized arrangements.

10–5. Illustrate the relationship of the structure and function of DNA to protein synthesis and the characteristics of an organism.

10–6. Explain that a unit of hereditary information is called a gene, and genes may occur in different forms called alleles (e.g., *gene for pea plant height has two alleles, tall and short*).

10–7. Describe that spontaneous changes in DNA are mutations, which are a source of genetic variation. When mutations occur in sex cells, they may be passed on to future generations; mutations that occur in body cells may affect the functioning of that cell or the organism in which that cell is found.

10–8. Use the concepts of Mendelian and non-Mendelian genetics (e.g., *segregation, independent assortment, dominant and recessive traits, sex-linked traits and jumping genes*) to explain inheritance.

10–9. Describe how matter cycles and energy flows through different levels of organization in living systems and between living systems and the physical environment. Explain how some energy is stored and much is dissipated into the environment as thermal energy (e.g., *food webs and energy pyramids*).

10–10. Describe how cells and organisms acquire and release energy (*photosynthesis, chemosynthesis, cellular respiration and fermentation*).

10–11. Explain that living organisms use matter and energy to synthesize a variety of organic molecules (e.g., *proteins, carbohydrates, lipids and nucleic acids*) and to drive life processes (e.g., *growth, reacting to the environment, reproduction and movement*).

10–12. Describe that biological classification represents how organisms are related, with species being the most fundamental unit of the classification system. Relate how biologists arrange organisms into a hierarchy of groups and subgroups based on similarities and differences that reflect their evolutionary relationships.

10–13. Explain that the variation of organisms within a species increases the likelihood that at least some members of a species will survive under gradually changing environmental conditions.

10–14. Relate diversity and adaptation to structures and their functions in living organisms (e.g., *adaptive radiation*).

10–15. Explain how living things interact with biotic and abiotic components of the environment (e.g., *predation, competition, natural disasters and weather*).

10–16. Relate how distribution and abundance of organisms and populations in ecosystems are limited by the ability of the ecosystem to recycle materials and the availability of matter, space and energy.

10–17. Conclude that ecosystems tend to have cyclic fluctuations around a state of approximate equilibrium that can change when climate changes, when one or more new species appear as a result of immigration or when one or more species disappear.

10–18. Describe ways that human activities can deliberately or inadvertently alter the equilibrium in ecosystems. Explain how changes in technology/biotechnology can cause significant changes, either positive or negative, in environmental quality and carrying capacity.

10–19. Illustrate how uses of resources at local, state, regional, national, and global levels have affected the quality of life (e.g., *energy production and sustainable vs. nonsustainable agriculture*).

10–20. Recognize that a change in gene frequency (*genetic composition*) in a population over time is a foundation of biological evolution.

10–21. Explain that natural selection provides the following mechanism for evolution; undirected variation in inherited characteristics exist within every species. These characteristics may give individuals an advantage or disadvantage compared to others in surviving and reproducing. The advantaged offspring are more likely to survive and reproduce. Therefore, the proportion of individuals that have advantageous characteristics will increase. When an environment changes, the survival value of some inherited characteristics may change.

10–22. Describe historical scientific developments that occurred in evolutionary thought (e.g., *Lamarck and Darwin, Mendelian Genetics and modern synthesis*).

10–23. Describe how scientists continue to investigate and critically analyze aspects of evolutionary theory.

10–24. Analyze how natural selection and other evolutionary mechanisms (e.g. *genetic drift, immigration, emigration, mutation*) and their consequences provide a scientific explanation for the diversity and unity of past life forms, as depicted in the fossil record, and present life forms.

10–25. Explain that life on Earth is thought to have begun as simple, one celled organisms approximately 4 billion years ago. During most of the history of Earth only single celled microorganisms existed, but once cells with nuclei developed about a billion years ago, increasingly complex multicellular organisms evolved.

10–26. Use historical examples to explain how new ideas are limited by the context in which they are conceived. These ideas are often rejected by the scientific establishment; sometimes spring from unexpected findings; and usually grow slowly through contributions from many different investigators (e.g., *biological evolution, germ theory, biotechnology and discovering germs).*

10–27. Describe advances in life sciences that have important long-lasting effects on science and society (e.g., *biological evolution, germ theory, biotechnology and discovering germs*).

10–28. Analyze and investigate emerging scientific issues (e.g., *genetically modified food, stem cell research, genetic research and cloning*).

PHYSICAL SCIENCES: PS

BENCHMARKS

PS:A Describe that matter is made of minute particles called atoms and atoms are comprised of even smaller components. Explain the structure and properties of atoms.

PS:B Explain how atoms react with each other to form other substances and how molecules react with each other or other atoms to form even different substances.

PS:C Describe the identifiable physical properties of substances (e.g., *color, hardness, conductivity, density, concentration and ductility*). Explain how changes in these properties can occur without changing the chemical nature of the substance.

PS:D Explain the movement of objects by applying Newton's three laws of motion.

PS:E Demonstrate that energy can be considered to be either kinetic (*motion*) or potential (*stored*).

PS:F Explain how energy may change form or be redistributed but the total quantity of energy is conserved.

PS:G Demonstrate that waves (e.g., *sound, seismic, water and light*) have energy and waves can transfer energy when they interact with matter.

PS:H Trace the historical development of scientific theories and ideas, and describe emerging issues in the study of physical sciences.

GRADE NINE GRADE LEVEL INDICATORS

9–1. Recognize that all atoms of the same element contain the same number of protons, and elements with the same number of protons may or may not have the same mass. Those with different masses (*different numbers of neutrons*) are called isotopes.

9–2. Illustrate that atoms with the same number of positively charged protons and negatively charged electrons are electrically neutral.

9–3. Describe radioactive substances as unstable nuclei that undergo random spontaneous nuclear decay emitting particles and / or high energy wavelike radiation.

9–4. Show that when elements are listed in order according to the number of protons (called the atomic number), the repeating patterns of physical and chemical properties identify families of elements. Recognize that the periodic table was formed as a result of the repeating pattern of electron configurations.

9–5. Describe how ions are formed when an atom or a group of atoms acquire an unbalanced charge by gaining or losing one or more electrons.

9–6. Explain that the electric force between the nucleus and the electrons hold an atom together. Relate that on a larger scale, electric forces hold solid and liquid materials together (e.g., *salt crystals and water*).

9–7. Show how atoms may be bonded together by losing, gaining or sharing electrons and that in a chemical reaction, the number, type of atoms and total mass must be the same before and after the reaction.

9–8. Demonstrate that the pH scale (0–14) is used to measure acidity and classify substances or solutions as acidic, basic, or neutral.

9–9. Investigate the properties of pure substances and mixtures (e.g., *density, conductivity, hardness, properties of alloys, superconductors and semiconductors*).

9–10. Compare the conductivity of different materials and explain the role of electrons in the ability to conduct electricity.

9–11. Explain how thermal energy exists in the random motion and vibrations of atoms and molecules. Recognize that the higher the temperature, the greater the average atomic or molecular motion, and during changes of state the temperature remains constant.

9–12. Explain how an object's kinetic energy depends on its mass and its speed ($KE = 1/2mv^2$).

9–13. Demonstrate that near Earth's surface an object's gravitational potential energy depends upon its weight (*mg where m is the object's mass and g is the acceleration due to gravity*) and height (*h*) above a reference surface ($PE = mgh$).

9–14. Summarize how nuclear reactions convert a small amount of matter into a large amount of energy. (*Fission involves the splitting of a large nucleus into smaller nuclei; fusion is the joining of two small nuclei into a larger nucleus at extremely high energies.*)

9–15. Trace the transformations of energy within a system (e.g., *chemical to electrical to mechanical*) and recognize that energy is conserved. Show that these transformations involve the release of some thermal energy.

9–16. Illustrate that chemical reactions are either endothermic or exothermic (e.g., *cold packs, hot packs and the burning of fossil fuels*).

9–17. Demonstrate that thermal energy can be transferred by conduction, convection or radiation (e.g., *through materials by the collision of particles, moving air masses or across empty space by forms of electromagnetic radiation*).

9–18. Demonstrate that electromagnetic radiation is a form of energy. Recognize that light acts as a wave. Show that visible light is a part of the electromagnetic spectrum (e.g., *radio waves, microwaves, infrared, visible light, ultraviolet, X-rays, and gamma rays*).

9–19. Show how the properties of a wave depend on the properties of the medium through which it travels. Recognize that electromagnetic waves can be propagated without a medium.

9–20. Describe how waves can superimpose on one another when propagated in the same medium. Analyze conditions in which waves can bend around corners, reflect off surfaces, are absorbed by materials they enter, and change direction and speed when entering a different material.

9–21. Demonstrate that motion is a measurable quantity that depends on the observer's frame of reference and describe the object's motion in terms of position, velocity, acceleration and time.

9–22. Demonstrate that any object does not accelerate (*remains at rest or maintains a constant speed and direction of motion*) unless an unbalanced (*net*) force acts on it.

9–23. Explain the change in motion (*acceleration*) of an object. Demonstrate that the acceleration is proportional to the net force acting on the object and inversely proportional to the mass of the object ($F_{net} = ma$). Note that weight is the gravitational force on a mass.)

9–24. Demonstrate that whenever one object exerts a force on another, an equal amount of force is exerted back on the first object.

9–25. Demonstrate the ways in which frictional forces constrain the motion of objects (e.g., *a car traveling around a curve, a block on an inclined plane, a person running, an airplane in flight*).

9–26. Use historical examples to explain how new ideas are limited by the context in which they are conceived; are often initially rejected by the scientific establishment; sometimes spring from unexpected findings; and usually grow slowly through contributions from many different investigators (e.g., *atomic theory, quantum theory and Newtonian mechanics*).

9–27. Describe advances and issues in physical science that have important, long-lasting effects on science and society (e.g., *atomic theory, quantum theory, Newtonian mechanics, nuclear energy, nanotechnology, plastics, ceramics and communication technology*).

GRADE TEN GRADE LEVEL INDICATORS

No Indicators present in this grade for Physical Sciences standard.

SCIENCE AND TECHNOLOGY: ST

BENCHMARKS

ST:A Explain the ways in which the processes of technological design respond to the needs of society.
ST:B Explain that science and technology are interdependent; each drives the other.

GRADE NINE GRADE LEVEL INDICATORS

9–1. Describe means of comparing the benefits with the risks of technology and how science can inform public policy.
9–2. Identify a problem or need, propose designs and choose among alternative solutions for the problem.
9–3. Explain why a design should be continually assessed and the ideas of the design should be tested, adapted and refined.

GRADE TEN GRADE LEVEL INDICATORS

10–1. Cite examples of ways that scientific inquiry is driven by the desire to understand the natural world and how technology is driven by the need to meet human needs and solve human problems.
10–2. Describe examples of scientific advances and emerging technologies and how they may impact society.
10–3. Explain that when evaluating a design for a device or process, thought should be given to how it will be manufactured, operated, maintained, replaced and disposed of in addition to who will sell, operate and take care of it. Explain how the costs associated with these considerations may introduce additional constraints on the design.

SCIENTIFIC INQUIRY: SI

BENCHMARKS

SI:A Participate in and apply the processes of scientific investigation to create models and to design, conduct, evaluate and communicate the results of these investigations.

GRADE NINE GRADE LEVEL INDICATORS

9–1. Distinguish between observations and inferences, given a scientific situation.
9–2. Research and apply appropriate safety precautions when designing and conducting scientific investigations (e.g., *OSHA, Material Safety Data Sheets [MSDS], eyewash, goggles and ventilation).*
9–3. Construct, interpret and apply physical and conceptual models that represent or explain systems, objects, events or concepts.
9–4. Decide what degree of precision based on the data is adequate and round off the results of calculator operations to the proper number of significant figures to reasonably reflect those of the inputs.
9–5. Develop oral and written presentations using clear language, accurate data, appropriate graphs, tables, maps and available technology.
9–6. Draw logical conclusions based on scientific knowledge and evidence from investigations.

GRADE TEN GRADE LEVEL INDICATORS

10–1. Research and apply appropriate safety precautions when designing and conducting scientific investigations (e.g. *OSHA, MSDS, eyewash, goggles and ventilation*).
10–2. Present scientific findings using clear language, accurate data, appropriate graphs, tables, maps and available technology.
10–3. Use mathematical models to predict and analyze natural phenomena.
10–4. Draw conclusions from inquiries based on scientific knowledge and principles, the use of logic and evidence (*data*) from investigations.
10–5. Explain how new scientific data can cause any existing scientific explanation to be supported, revised or rejected.

SCIENTIFIC WAYS OF KNOWING: SW

BENCHMARKS

SW:A Explain that scientific knowledge must be based on evidence, be predictive, logical, subject to modification and limited to the natural world.

SW:B Explain how scientific inquiry is guided by knowledge, observations, ideas and questions.

SW:C Describe the ethical practices and guidelines in which science operates.

SW:D Recognize that scientific literacy is part of being a knowledgeable citizen.

GRADE NINE GRADE LEVEL INDICATORS

9–1. Comprehend that many scientific investigations require the contributions of women and men from different disciplines in and out of science. These people study different topics, use different techniques and have different standards of evidence but share a common purpose — to better understand a portion of our universe.

9–2. Illustrate that the methods and procedures used to obtain evidence must be clearly reported to enhance opportunities for further investigations.

9–3. Demonstrate that reliable scientific evidence improves the ability of scientists to offer accurate predictions.

9–4. Explain how support of ethical practices in science (e.g., *individual observations and confirmations, accurate reporting, peer review and publication)* are required to reduce bias.

9–5. Justify that scientific theories are explanations of large bodies of information and/or observations that withstand repeated testing.

9–6. Explain that inquiry fuels observation and experimentation that produce data that are the foundation of scientific disciplines. Theories are explanations of these data.

9–7. Recognize that scientific knowledge and explanations have changed over time, almost always building on earlier knowledge.

9–8. Illustrate that much can be learned about the internal workings of science and the nature of science from the study of scientists, their daily work and their efforts to advance scientific knowledge in their area of study.

9–9. Investigate how the knowledge, skills and interests learned in science classes apply to the careers students plan to pursue.

GRADE TEN GRADE LEVEL INDICATORS

10–1. Discuss science as a dynamic body of knowledge that can lead to the development of entirely new disciplines.

10–2. Describe that scientists may disagree about explanations of phenomena, about interpretation of data or about the value of rival theories, but they do agree that questioning, response to criticism and open communication are integral to the process of science.

10–3. Recognize that science is a systematic method of continuing investigation, based on observation, hypothesis testing, measurement, experimentation, and theory building, which leads to more adequate explanations of natural phenomena.

10–4. Recognize that ethical considerations limit what scientists can do.

10–5. Recognize that research involving voluntary human subjects should be conducted only with the informed consent of the subjects and follow rigid guidelines and/or laws.

10–6. Recognize that animal-based research must be conducted according to currently accepted professional standards and laws.

10–7. Investigate how the knowledge, skills and interests learned in science classes apply to the careers students plan to pursue.

INDEX

A

Abiotic components, 181, 182, 187, 190, 207
Acceleration, 124, 126, 130, 207
Acid, 118, 121, 207
Acid rain, 188, 190
Action forces, 128
Active transport, 153, 158, 207
Adaptive radiation, 175, 207
Adenosine triposphate (ATP), 149, 152, 154, 207
Alkali metals, 115
Alleles, 162, 167, 207
Alternative energy sources, 189
Animal cells, 152
Aquatic ecosystem, 182, 190
Archaebacteria, 173–174, 207
Asthenosphere, 75
Atmosphere, 82, 84, 85, 207
Atom, 98–101, 104, 207
Atomic number, 100, 106, 110, 207
Atomic symbol, 100, 106

B

Base, 118, 207
"Big bang" theory, 68–69, 70, 207
Binary fission, 155, 207
Biodiversity, 189, 190, 207
Biome, 83, 84, 85, 207
Biotechnology, 166, 207
Biotic components, 181, 183, 187, 190, 207
Boiling point, 117, 121, 207

C

Carnivores, 185, 190
Carrying capacity, 184
Cell, 149–155, 207
Cell membrane, 155, 151, 158, 207
Cell theory, 150, 158, 207
Cell wall, 152, 207
Cellular division, 155–157
Cellular processes, 153–155, 158
Cellular respiration, 154, 158, 207
Centrifugal force, 80
Chain reaction, 143
Chemical energy, 135
Chemical equations, 104, 106
Chemical property, 112, 207
Chemical reaction, 104, 106, 120, 207
Chromatid pairs, 157
Chromosome, 155, 207
Classification, 176, 207
Commensalism, 184, 207
Competition, 184, 190
Compound, 104, 208
Conclusions, 48–49
Condensation, 80
Conduction, 119, 137, 143, 208
Conservation, 189
Conservation of energy, 105, 139, 208
Conservation of mass, 105, 106
Conservation of momentum, 129
Continental crust, 73
Continental drift, 74
Control group, 45, 51, 208
Convection, 67, 75, 137, 143, 208
Convergent plate boundaries, 76, 208
Copernicus, Nicolaus, 65
Core, 74, 85, 208
Cosmic background radiation, 68
Covalent bond, 102–103, 105, 106, 208
Crick, Francis, 165, 167
Crust, 73, 85, 208
Cytoplasm, 150, 151, 158, 208

D

Darwin, Charles, 170, 171–172, 178
Decomposers, 185, 208
Density, 119, 208
Deoxyribonucleic acid (DNA), 60, 149, 150, 155, 165–167, 173, 208
Diffraction, 141, 143
Displacement, 125, 130
Distance, 125, 130
Dominant trait, 162, 167, 208
Doppler Effect, 141

E

Ecology, 181–186
Ecosystem, 181, 208
Einstein, Albert, 36
Electricity, 119, 136, 208
Electromagnetic radiation, 133, 142, 143, 208
Electromagnetic spectrum, 141–142
Electromagnetic waves, 140
Electron, 99, 101, 106, 114, 208
Element, 98, 104, 111, 114, 208
Endocytosis, 153
Endoparasites, 183
Endoplasmic reticulum, 151, 158, 208
Endothermic energy, 138, 208
Energy, 133, 143, 208
Energy levels, 99, 102, 105
Equilibrium, 184, 208
Eukaryotic cells, 149, 151, 155, 158, 208
Evolution, 170–174
Exocystosis, 154
Exothermic energy, 138, 208
Experiment, 33, 208
Experimental design, 44
Experimental group, 45, 51
Extended-response questions, 25–30

F

Fault, 78
Fermentation, 155, 158, 209
Floating, 120
Folding, 76, 78, 209
Food chain,186, 209
Force, 133, 209
Fossil, 91, 172, 178, 209
Frequency, 140
Friction, 124, 129, 130

G

Gamete, 163
Gamma rays, 142
Gases, 117, 121
Gene, 162, 209
Gene frequency, 174
Genetic drift, 175, 178, 209
Genetic mutation, 172
Genotype, 162, 167, 209
Global warming, 93, 94, 188, 190
Golgi complex, 152, 158, 209
Granite intrusion, 91
Gravitational potential energy, 133, 135, 142, 143, 209
Gravity, 66, 70, 75, 209
Great Lakes ecosystem, 187
Greenhouse effect, 93, 94, 188
Groups, 110, 113, 120

H

Heliocentric theory, 65, 70, 209
Herbivores, 185, 190
Heterogeneous mixtures, 111–112
Heterotrophs, 185
Heterozygous, 162, 167, 209
Homeostasis, 153, 158, 209
Homogeneous mixtures, 111–112
Homologous chromosomes, 157
Homozygous, 162, 167

Human Genome Project, 166
Hydrosphere, 79, 84, 85, 209
Hypothesis, 44, 50, 209

I, J, K

Igneous rock, 79, 209
Incomplete dominance, 163
Inertia, 125
Ion, 101–102, 106, 209
Ionic bonding, 102–103, 105, 106
Isotope, 100, 101, 106, 209
Kinetic energy, 133, 134–135, 139, 142, 143, 209
Kingdom, 177, 178

L

Lamarck, Jean Baptiste de, 171, 178
Law of Independent Assortment, 162, 167
Law of Inertia, 125, 130
Law of Superposition, 91, 94
Liquids, 117, 121
Lithosphere, 75, 84, 85, 209
Litmus paper, 118
Lyell, Charles, 90
Lysosomes, 152, 158

M

Magma, 76, 209
Magnetic striping, 78
Mantle, 74, 85, 209
Mass, 98, 105, 210
Material Safety Data Sheets (MSDS), 47
Mathematics, role of, 33
Mating, 175, 178
Matter, 98, 110, 112, 209
Measurements, taking, 47–48
Mechanical energy, 138, 139
Mechanical waves, 140, 143
Meiosis, 156–157, 158, 209
Mendel, Gregor, 161–162, 167
Mendeleev, Dmitri, 113
Metal, 114–115, 116, 209
Metalloids, 115, 116
Metamorphic rock, 79
Mid-Atlantic ridge, 76
Mitochondria, 152, 158, 209
Mitosis, 155–156, 157, 209
Mixture, 110, 111, 120, 209
Molecule, 103, 106, 209
Momentum, 124, 129, 130
Mutation, 166
Mutualism, 184, 190, 209

N

Neutron, 99, 101, 106, 210
Newton, Sir Isaac, 35, 66
Newton's First Law, 124, 125, 130, 210
Newton's Second Law, 124, 126, 127, 130, 210
Newton's Third Law, 124, 128, 129, 130, 210
Noble gases, 115, 116
Nonmetals, 115, 116
Nuclear energy, 136, 143
Nuclear fission, 136, 143
Nuclear fusion, 67, 136, 143
Nucleus, 151, 158

O, P

Omnivore, 185, 190
Organelle, 151, 210
Ozone layer, 188, 190
Parasitism, 183, 190, 210
Passive transport, 153, 158
Pasteur, Louis, 35
Periodic Table of Elements, 100, 110, 113–114, 210
Periods, 92, 110, 113, 120
pH scale, 110, 118–119, 121, 210
Phenotype, 162, 167
Photons, 68, 141
Photosynthesis, 149, 152, 154, 158, 210
Plate boundaries, types of, 76–77
Plate tectonic theory, 75–78
Potential energy, 135
Precipitation, 79–80
Predation, 183, 190, 210
Prokarytoic cells, 149, 150, 155, 158, 210
Proton, 99, 101, 106, 210
Punnett square, 162–163

R

Radiation, 67, 137, 143, 211
Radioactive, 92, 101, 106
Radiometric dating, 92, 94, 211
Reactants, 104, 211
Reaction forces, 128
Recessive trait, 162, 167, 211
Recombinant DNA, 166
Reflection, 141, 143, 211
Refraction, 141, 143, 211
Ribosomes, 151, 158
Richter Scale, 78
Risk vs. benefit, 59–60
RNA, 166
Rock formations, 79
Rocks, age of, 91–92

S

Safety considerations, 46, 51
Scientific inquiry, 33, 55–57
Scientific investigation, 43–50, 51, 211
Sedimentary rock, 79, 91, 211
Seismic waves, 73, 211
Seismograph, 78
Self-pollination, 161
Semi-conductors, 115, 116, 119
Sex-linked traits, 164
Short-answer questions, 25–30
Solid, 116, 121, 211
Speciation, 175, 178, 211
Species, 177, 178, 211
Splice genes, 166
Star, 67, 70, 211
Subatomic particles, 101
Subduction, 75
Subduction zone, 77, 211
Substance, 110, 120, 211
Superconductors, 119
Supernova, 67

T

Tables, 13–14
Technology, 55–58, 211
Temperate deciduous forests, 83
Temperature changes, 117, 134, 211
Terrestrial ecosystem, 182, 190
Theory, 33, 36, 211
Theory of natural selection, 171, 178
Theory of Universal Gravitation, 36
Thermal energy, 133, 134, 137, 142, 143, 211
Tide, 80, 211
Timeline, 92
Tornado alleys, 82
Tornadoes, 82
Transform plate boundaries, 77
Transmutation, 101
Transpiration, 79
Trophic level, 186
Tropical hurricanes, 82
Tropical rain forests, 83
Troposphere, 82

U, V, W, X, Y, Z

Ultraviolet light, 142
Uniform motion, 126
Variable, 44–45, 50, 211
Velocity, 124, 125, 130, 211
Vertebrates, 173
Vestigial structures, 172, 178
Water cycle, 79–80, 85, 211
Watson, James, 165, 167
Wave, 81, 133, 143, 211
Wave length, 140
Wegener, Alfred, 74
X-rays, 142